CLIMATE BOGEYMAN

The Criminal Insanity of the Global Warming / Climate Change Hoax

By M. S. KING / © 2017

"History is indeed little more than the register of the crimes, follies and misfortunes of mankind."

Edward Gibbon, English historian (1737-1794)

From: *"The Decline and Fall of the Roman Empire"*

About the Author

M. S. King, the webmaster of **TomatoBubble.com**, is a private investigative journalist and researcher based in the New York City area. A 1987 graduate of Rutgers University, King's subsequent 30 year career in Marketing & Advertising has equipped him with a unique perspective when it comes to understanding how "public opinion" is indeed scientifically manufactured.

Madison Ave marketing acumen combines with 'City Boy' instincts to make M.S. King one of the most tenacious detectors of "things that don't add up" in the world today. Says King of his admitted quirks, irreverent disdain for "conventional wisdom," and uncanny ability to ferret out and weave together important data points that others miss: *"Had Sherlock Holmes been an actual historical personage, I would have been his reincarnation."*

King is also the author of *The Bad War, The War Against Putin, The British Mad Dog, Planet Rothschild, The Real Roosevelts, Napoleon vs the Old & New World Orders, I Don't Like Ike* and more. King's other interests include: the animal kingdom, philosophy, chess, cooking, literature and history *(with emphasis on events of the late 19th through the 20th centuries).*

Amazon Author Page: <u>M S King</u>

TomatoBubble.com

TABLE OF CONTENTS

CLIMATE BOGEYMAN

The Criminal Insanity of the Global Warming / Climate Change Hoax

*GW / CC = Global Warming / Climate Change

INTRODUCTION

As we all know, there is a group of scientists, journalists, and politicians that refer to catastrophic manmade CO_2-induced Global Warming / Climate Change (GW/CC) as the *"greatest crisis facing humanity today."* On the other end of the opinion spectrum, another lesser known group of scientists, journalists, and politicians refer to manmade GW/CC as *"the greatest lie ever told."* The opposing camps, as well as those "moderate" types who will always gravitate to the mushy "middle ground," cannot all be right.

So, how are we mere mortals supposed to decide which group of all-mighty scientists and "intellectuals" to trust? The answer: believe no one! Trust *your own* logic and reason to evaluate only the hard facts and nothing else, and then decide on your own. You'll be amazed at the revelations that come to you when you make the commitment to do the hard work of thinking independently of the "experts."

As you have no doubt already inferred from the title of this book, your free-thinking author here, in proud and certain defiance of "the great and the good" who continue to proclaim GW/CC to be "settled science," and after much truly objective research and logical analysis, has aligned himself firmly in the camp of the "climate deniers," and is hopeful of helping readers to also move to that side.

What separates this work from that of so many other excellent books and documentaries which debunk manmade GW/CC is that it provides a *complete picture* of the grand hoax – one that not only encompasses the Fake Science of GW/CC, but also ties it into the public relations component and the historical geopolitical context which drives the "**Climate Bogeyman.**"

It is self-defeating to "respectfully disagree" with the warmists. Any respect afforded to that crowd implies that they, in the words of one "denialist" film maker, are "not evil, just wrong." To hell with pulling punches! These people need to be called out. Now there are indeed many innocent fools and dupes, including some scientists, who, under the influence of propaganda and "expert" authority, have obediently swallowed the dogmatic dung of GW/CC and actually do believe in it. As for the inner circle movers and shakers behind the great scam; these EVIL conspirators know *exactly* what they are doing and they need to be placed on the defensive and called out not as fools or merely sloppy practitioners of science, but as the criminal hoaxsters that they and their invisible handlers truly are.

Let's unmask this gigantic fraud.

WHAT IS GLOBAL WARMING / CLIMATE CHANGE (GW/CC)?

Scary science --- or science fiction?

The "Greenhouse Effect" aka "Global Warming" aka "Climate Change" refers to the warming that happens when certain gases in Earth's atmosphere trap heat. These gases, mostly emitted naturally, allow in the sunlight light but hinder heat from escaping, like the glass walls of a greenhouse.

First, sunlight shines onto the Earth's surface, where it gets absorbed and then radiates back into the atmosphere as heat. In the atmosphere, "greenhouse gas" molecules trap some of this heat because they are not transparent to some wavelengths of thermal radiation. When greenhouse gases absorb thermal infrared energy; their temperature rises. The rest escapes into space. The more greenhouse gases are in the atmosphere, the more heat is retained.

Carbon Dioxide (CO_2), along with water vapor (H_2O), methane (CH_4) and nitrous oxide (N_2O), are considered to be "greenhouse gases." CO_2 is both produced and absorbed by various natural sources and essential for plants to grow. But because industrial and automotive human activity *(cars, airplanes, smokestacks etc)* also emit CO_2, it is claimed that the Earth's natural mechanisms cannot cycle through or balance out all of this extra CO_2.

It is said that if manmade CO_2, and, to a lesser extent, manmade N_2O and CH_4 are not kept under strict control by government, the emissions will cause so much heat to be trapped "inside the greenhouse" that the resulting temperature increases will melt enough land-based polar ice from Antarctica and Greenland to flood the world's coastal areas in a catastrophic way. Global Warming / Climate Change (GW/CC) is also expected to increase the number of violent storms and extreme temperature swings – both hot *and* cold.

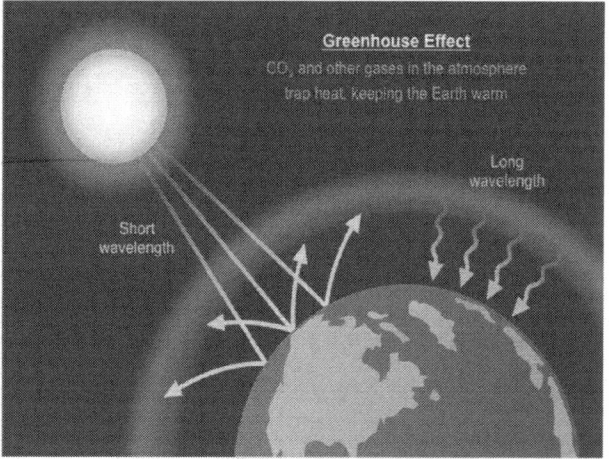

SECTION 1
THE FAKE NEWS OF GLOBAL WARMING / CLIMATE CHANGE

"I am not a scientist, but I don't need to be. Because the world's scientific community has spoken, and they have given us our prognosis, if we do not act together, we will surely perish." (1)

- Leonardo DiCaprio / Actor

COMMENTARY

We have arrived at a point in time in which a slight majority of the public now believes that at least some degree of manmade Global Warming / Climate Change (GW/CC) -- and at least some degree of manmade sea-level rise *(caused by melting polar ice)* is also happening.

That 55-60% of the public that has fallen for this hoax *(and we will explain why it is indeed a hoax in the section on Fake Science)* can further be sub-divided into one part that doesn't really foresee catastrophic consequences and hence, isn't too worried about it – and another part of roughly equal size which believes that great calamity will befall the next generation if we do no act to "save the planet" from overdosing on CO_2.

Because manufactured public opinion is used to support public policy, it is very important to understand how we have arrived at this disturbing degree of mass madness. It was not "science," fake or real, which created this public misconception. Though a small handful of self-styled students of "science" may indeed have been convinced by the elaborate math tricks built upon flawed data; it was mainly Fake News, not Fake Science, which led 90% + of the believers in GW/CC to form an erroneous opinion.

This section will review the main opinion-shaping events and developments of the past four decades. It was these factors, each amplified by overwhelming "mainstream media" coverage, which, slowly but surely over these past 40 years, moved the public polling numbers.

1970
THE FIRST ANNUAL "EARTH DAY"

Earth Day is a worldwide annual event that takes place every April 22nd. The event was conceived by liberal activist **John McConnell** in 1969 at a United Nations Conference in San Francisco. A radical Communist "hippie" and convicted murderer named **Ira Einhorn** claims that he was a co-founder of Earth Day. Although liberals insist that Einhorn's claim is not true, he was indeed one of the main organizers for the Philadelphia events of Earth Day 1970.

Another of the key players *(the players which we are able to see, that is)* behind the launch of Earth Day was liberal **Senator Gaylord Nelson** (D-WI), who was later recognized for his efforts with a Presidential Medal of Freedom -- awarded to him by President **Bill Clinton**.

Earth Day 1970 was kicked-off with unusual fanfare and media hype, including front page spreads in virtually every newspaper in America, and a CBS TV special hosted by the legendary anchorman, **Walter Cronkite**, aka *"The Most Trusted Man in America."* In addition to major events being held in New York and Einhorn's Philadelphia, celebrations took place at two thousand colleges and universities, roughly ten thousand primary and secondary schools, and hundreds of communities across America.

It was obvious that America's ruling class, through its wholly-owned media, was strongly pushing the new "holiday." We will address the true motive for the hype later on, in Section 5. But as for the purported motive -- the "cause" which drove so many well-meaning but ignorant people into the streets -- it was all about "saving the planet" from oil spills, air pollution, water pollution and various yet to be discovered *(or invented)* crises.

The artificially whipped-up "public opinion" which grew out of the event led to the establishment of the **EPA *(Environmental Protection Agency)*** less than 8 months later. Like all new government agencies, the EPA started off small and with limited powers. Today, nearly a half-century later, the intrusive EPA is the most dictatorial and economically destructive of all of America's regulatory agencies. The out-of-control agency has even gone so far as to declare CO_2 *(plant food)* a "pollutant." Yes indeed. The Communist murderer and claimed Earth Day co-founder Ira Einhorn must be smiling from his jail cell.

Walter Cronkite *and the media's heavy promotion of "Earth Day 1970"*
was a sure indication of a Globalist push for some sinister agenda.

1980's:
"THE GREENHOUSE EFFECT / GLOBAL WARMING IS COMING!"

With the passing of some colder-than-normal winters in northeast America, the attempt to concoct a manmade ice age scare fizzled out by the end of the 1970's *(more on that, later)*. You see, the difficulty with selling an ice-age hoax is that it requires an observable year-by-year southern expansion of permanent ice coverage in inhabited areas. That's essentially what an ice age is. People living in the north would *have to* notice that their winter snow accumulation suddenly wasn't melting until the summertime, if at all. Otherwise, the ice age scare won't fly.

At a time when world population and CO_2 *(plant-food)* emissions were far lower than they are today, concerns about an ice-age were suddenly replaced with tales of a "Greenhouse Effect" expected to wreak havoc by the Year 2000. After nearly 40 years, none of the dire predictions have come to pass as "doomsday" is again and again pushed up to yet another far off day in "our children's future."

By creating the scam of a CO_2-caused "Greenhouse Effect" that locks in heat so that it can't escape into the upper atmosphere, the high-level controllers of the "Fake News" could deceive the media-worshipping public into believing that polar ice would one day melt away due to the "trapped heat" -- thus flooding our coastlines as temperatures rise. Since nobody actually lives in the interior of

14

Greenland or Antarctica, we must take the "scientists" word for it that the polar ice is gradually "melting" and that the problem will rapidly accelerate in the future.

1980's / Scare Stories about the "Greenhouse Effect / Global Warming" Spread Quickly

1980's- PRESENT
THE MEDIA OPERATES UNDER STRICT DIRECTIVES TO NEVER INTERVIEW SCIENTISTS WHO DENY GW/CC

Throughout so many large and "prestigious" institutions of the "mainstream media" of the US and Europe, there has been a total absence of opposing voices on the issue of GW/CC. A thorough search through the archives of the all-mighty **New York Times** and the **Washington Post** will turn up claims by all the major warmists *(such as Stephen Schneider, Michael Mann, James Hansen et al.);* but never, and I mean *never*, any corrective counter claims by the most prominent "climate deniers" *(such as Richard Lindzen, John Coleman, Fred Singer et al)*.

If the "deniers" are ever interviewed at all, it is always in a dismissive "straw man" sort of way with a hostile and rhetorically skilled interviewer tightly controlling the direction of the conversation. But never are the sound arguments of the "deniers" ever permitted to be presented in full, and without adversarial interruption.

The major news networks *(ABC, NBC, CBS, PBS, CNN)* also give the "deniers" the total silent treatment, as do the major print and online news magazines *(Time, Newsweek, US News & World Report, the Economist, the Huffington Post etc.)* and science/nature magazines *(Popular Science, Nature, Scientific American, National Geographic, Discover etc)*. Even the allegedly "conservative" FOX News and Wall Street Journal, though they will allow a general questioning of the cause and extent of the "problem," will only give limited coverage to scientists who deny man-made GW/CC altogether.

The most blatant example of censorship was the last minute decision of PBS to stop the excellent British documentary ***The Greenhouse Conspiracy*** from airing on an affiliate station in 1991. Said PBS producer Linda Harrar of her decision to kill the film:

"I'm not sure it is useful to include every single point of view." **(2)**

How can it be that clownish High School science teachers such as **Bill Nye** and **Neil DeGrasse Tyson** get interviewed as "authorities" about GW/CC, while serious men of real science, men considered among the elite in their fields, are ignored? How is that **PBS** and the **National Geographic Channel** can air "crockumentaries" on GW/CC, but have never *once* broadcast any of a number of professionally-made films which challenge the official dogma? The media has a moral responsibility to the public to present all sides of a given issue. Who the bloody hell are they to decide what you should and should not see?

This blatant selectivity constitutes clear, convincing and undeniable evidence that the big media bosses are in collusion to deny the "deniers" access to the public. There can be no other logical inference than that.

BANNED FROM THE MAINSTREAM MEDIA!

*Dissenting climate scientists like **Richard Lindzen** (MIT), **Fred Singer** (University of Virginia), **John Coleman** (Weather Channel Founder) will occasionally pop up on small market cable news shows, but are NEVER allowed to come near any of America's major media outlets.*

1987
THE MONTREAL PROTOCOL ADDRESSES "THE HOLE IN THE OZONE LAYER"

Years before hysteria over "Global Warming" became the dominant environmental cause-of-the-day, the more urgent crisis was all about the fictitious "hole in the ozone layer," said to be caused by chlorofluorocarbons *(CFCs)* found in aerosol spray cans and refrigerants. The "hole," we were told, was letting in dangerous radiation harmful to life. **The Montreal Protocol** agreed to restrict the chemicals that allegedly damage the ozone layer.

The hoax of ozone layer depletion is not the subject of this expose and we do not wish to digress from our main topic. We raise this issue only because it is truly fascinating, in light of the hysteria which once surrounded this "crisis," how no one really ever talks about it anymore. Just like the 1970's ice-age mini-scare *(more on that, later)*, ozone hysteria seemed to suddenly subside when "The Greenhouse Effect" took center stage.

Evidently, the high level planners were, at varying times, not quite sure which environmental hoax to bet all of their chips on. When they finally settled on and chose to run exclusively with "The Greenhouse Effect / Global Warming," all the scary media talk about "the hole in the ozone layer" suddenly stopped, as if somebody flicked an "off" switch. Very strange!

Hysteria over spray-can-induced "ozone-holes" and killer radiation was also a big deal back in the 80's and 90's. It must have been put to rest so as not to distract from "Global Warming."

Remember the "Acid Rain" bogeyman from the 1980's?

Just like the Ozone Hole Scare, the hysteria about "Acid Rain" also disappeared suddenly.

1988
THE UN's IPCC IS ESTABLISHED

The **United Nations Intergovernmental Panel on Climate Change (IPCC)** was formed to collect and assess evidence on "Global Warming." Many of the most influential high priests of the cult of "Global Warming" sit on or advise this panel, and get paid very well to do so. The "prestigious" UN's obsession with this issue not only gives added credibility to the "crisis," but enables the world body to buy support for "saving the planet" from poorer undeveloped countries in exchange for getting aid packages in lieu of industrializing. IPCC is one of the driving forces behind promoting the hoax, and finding "solutions."

The UN's formidable clout and little-known influence over global public education curricula adds credibility to the false claims of the warmists.

MARCH, 1989
THE EXXON VALDEZ OIL SPILL IS USED TO VILIFY OIL DRILING

Although this particular event had nothing to do with the propaganda campaign to sell GW/CC, it was indeed used to attack and further control the oil industry – always one of the main hidden objectives of the warmists. For that reason, it ranks as a significant milestone of the very same Fake News propaganda campaign that encompasses GW/CC.

The *Exxon Valdez* oil spill occurred in Prince William Sound, Alaska when the giant tanker struck a reef and spilled 10.8 million gallons of crude oil over the course of a few days. It was, at the time,

20

the worst oil spill in history. Prince William Sound's remote location made response efforts difficult as the oil spill spread over 100's of miles of Alaska coastline and up to 10,000 square miles of ocean.

Night after night, *for weeks*, TV viewers were subjected to the ugly, "heartbreaking" sight of seals, otters and birds covered in muck. The coverage of the investigation of the spill then lasted for months. Certainly, the story was newsworthy. The problem was that the extended doom and gloom coverage of an event in which nobody was killed, and the public whipping of Exxon and the ship's captain, **Joseph Hazelwood**, were grossly disproportionate and unfair. As it is with automobile fatalities and airline crashes, accidents happen. Enough already!

The fact that nature's own natural cleansing mechanisms could, and did, restore the region to pristine condition within several years was never reported. Viewers were left with the false impression that a chunk of Alaska was gone forever.

The exaggerated and scary coverage of the event dealt a huge and lasting blow to the reputation of the oil industry – a blow that surely must have pleased the ultimate planners behind the GW/CC hoax. Coming just 1 year before the big 20th anniversary of "Earth Day," the *Exxon Valdez* spill gave the green activists and their hidden masters another reason to shout: *"See. We told you that we have to wean ourselves off from dirty oil."*

Indeed, warmists are *still* talking about *Exxon Valdez* and still lying about how the region "has never fully recovered." The related litigation against Exxon didn't end until 2016.

The warmists used the sad images of oil-covered birds and otters to relentlessly agitate against oil. The story's powerful imagery was kept alive by the media for months. Even today, in spite of the fact that the area is as clean as it was before the spill, they still bring up the event. Why?

APRIL, 1990
THE 20TH ANNIVERSARY OF "EARTH DAY" GOES INTERNATIONAL

The media-hype machine went into high gear to artificially promote the 20th Anniversary of Earth Day. Try as they might, the "holiday" just never really caught on with the great masses of America. However, among the political Left *(people who believe in big government and international government)* Earth Day 1990 was a big event.

It was an activist and government-connected leftist named **Dennis Hayes** who organized the big campaign and made the 20th Earth Day a global event. An estimated 750,000 people turned out in New York City's Central Park *(at least that's what organizers claim)*. In Washington DC, 100,000 - 250,000 *(depending on whose numbers you want to believe)* showed up. Across America and worldwide, many millions of other leftists and well-meaning liberal fools mobilized in 141 different countries to create global "awareness" of environmental issues.

Unlike Earth Day 1970, held at a time when the first theories of global *cooling* were being hatched, Earth Day 1990 focused heavily on, and gave a huge international boost to, the "Greenhouse Effect / Global Warming" hoax. Out of the hype over Earth Day 1990 came the planned push for a big conference which would materialize in the form of the **1992 United Nations Earth Summit** in Rio de Janeiro.

Deluded mobs of media-manipulated fools all over the world turned out to "save the planet."

1990 - 1996
"CAPTAIN PLANET" TARGETS CHILDREN WITH "GREENHOUSE EFFECT" PROPAGANDA

The children's TV show, **Captain Planet and the Planeteers** was broadcast by Globalist *(a person who wants one-world government)* **Ted Turner's** TBS network from 1990 to 1992. A sequel series, **The New Adventures of Captain Planet** was later broadcast from 1993-1996. Millions of vulnerable children, now in their 30's, were thus brainwashed at a young age by frightening cartoons which featured the green-haired **Captain Planet** and his young "Planeteers" battling against evil business and oil "eco-villains" such as "**Hoggish Greedly.**"

In one episode, Greedly, a monstrous-looking villain, plots to deliberately burn as much coal and oil as he can in order to melt the polar ice and flood the world. Only Captain Planet and the Planeteers can stop him. Here is some of the scary dialogue from that episode:

Greedly: *We'll burn up the Earth until this place (the North Pole) is as hot as the tropics. We'll burn all the coal and oil we can find.*

Greedly's talking computer associate: *Permit me to show you what will happen if you go and do that. Your burning will release tons of carbon dioxide into the air. The sunlight that would normally bounce back into space as heat will be trapped by the increased carbon dioxide. It's like rolling up the windows on a car parked in the sun, or like a greenhouse. After a few years, this global warming will melt the polar ice, raising the sea-level all over the world, changing weather and making storms worse.* **(3)**

(cue scary music)

Fortunately for the world, **Captain Planet** was able to thwart Greedly's conspiracy and "save the planet" -- at least for now. As with all captain Planet episodes, the green-haired goofball closed the show with a sanctimonious little speech.

Captain Planet: *Around the world, more (carbon dioxide) is added every day. That's why it's so important to burn less of those fuels that pollute the atmosphere and make earth hotter... We must stop this. The power is yours!* **(4)**

Any innocent child who watched this crap was influenced -- even traumatized -- by these blatant lies *(which we will analyze later on)*. Captain Planet action figures and a board game by University Games were also marketed to kids. Though the "super-hero" never caught on to the extent of a "Spider Man" or "Super Man," -- millions of TV-addicted children were indeed corrupted. And they walk among us today as adults with a hateful bias towards business, industry and the free market, and a worshipful bias in favor of big government and international bodies like the UN -- exactly as Ted Turner, who has donated **$1 BILLION** to the UN, and his Globalist crowd had planned.

And so it came to pass that millions of gullible children, at a time when they lacked the emotional and intellectual capacity to resist such clever brainwashing, became pre-disposed to believing this nonsense as adults.

1. In the early 1980's, images of the Statue of Liberty, predicted to be under water by 2000, were used to frighten school children across America.
*2. **Captain Planet and the Planeteers**: Propaganda aimed at helpless children -- many of whom have since grown up into adult warmists.*

Ted Turner, Captain Planet, Hoggish Greedly

24

1992
THE RIO "EARTH SUMMIT"

The environmentalist propaganda of the past two decades provided Globalist movers and shakers with the necessary cover needed to hold an **'Earth Summit'** in Rio de Janeiro, Brazil -- to "save the planet," of course *(rolling eyes)*. Just the fact that delegations from 172 countries attended this hyped-up conference served to convince many millions of more people that there must really be some truth to this "Greenhouse Effect." Why else would officials from so many countries be attending such an event?

Though no major policy changes came out of the "Earth Summit," it did achieve the framework for future protocols and agreements which we see coming together today -- such as the **"Agenda 21"** scheme which will undermine property rights and transfer enormous power to centralized governments and international institutions such as the United Nations. We will dig deeper into the question of motives in Section 5 of this book. For now, the important takeaway point about the Rio Earth Summit is that the big show gave another huge public propaganda boost to the promoters of the "Greenhouse Effect" -- soon to be referred to as "Global Warming."

The high and the mighty, and the great and the good all convened in Rio to "save the planet" from Global Warming and other environmental "crises" --- while the Establishment media hyped the event in print and on TV.

1992
SENATOR AL GORE PUBLISHES "EARTH IN THE BALANCE"

As the Earth Summit ended in Rio, on June 22, 1992, US Senator and soon-to-be Vice President, **Al Gore** (D-TN) published *Earth in the Balance: Ecology and the Human Spirit,* a book in which he called for a *"Global Marshall Plan (*a post WW 2 scheme to rebuild Europe) to "save the planet."* The highly alarmist book was hyped to the stars by the New York Times and the rest of the Piranha Press -- earning the already wealthy Senator a TON of royalties. Because the release coincided with the Presidential election season, Gore *(Bill Clinton's soon-to-be pick for his Vice-Presidential running mate)* and his error-ridden book were showered with enormous publicity. Though the book deals with a variety of topics, "Global Warming" is prominently featured.

Gore's important contribution to advancing the long march of this hoaxery lies in the fact that he became the highest and most visible public official to put his name so strongly behind the "fact" of "Global Warming." Tall, handsome, articulate and hyped to the stars by Hollywood and the fawning Fake News, **Bill Clinton's** ever-present sidekick sold the scam better and to a wider audience than anyone else had done before. Thanks to Gore and the powers propping him up, belief in the "crisis" of "Global Warming" was no longer exclusively limited to aging hippies and leftist college kids. With a "respected" political leader such as the "moderate" Al Gore so firmly behind it, many other Americans began leaning towards belief.

The propaganda build-up of the "intellectual" Al Gore and his pet issue of "Global Warming" went hand in hand.

The **Kyoto Protocol** is an international treaty which extends the 1992 **United Nations Framework Convention on Climate Change (UNFCCC)** that was worked up in Rio five years earlier. Just like Rio 1992, the more subdued Kyoto event, *(held in Kyoto, Japan)* was also promoted by the media, though not as intensely.

Kyoto committed nations to reduce CO_2 emissions, based on an alleged "scientific consensus" that a) "Global Warming" was occurring and b) it was likely that human-made CO_2 was causing it. The Kyoto Protocol was adopted in December, 1997 and became international law for countries still in it in February, 2005. Though US President George Bush essentially withdrew the US from the voluntary Protocol in 2001, future President Obama would once again work towards its objectives. Slowly but surely, these international conferences would move the world closer and closer toward the ultimate goal, which has nothing to do with "saving the planet" -- more on that subject, in Section 5.

1. US Vice Pres. Al Gore delivers the opening speech of the conference in Kyōto **2.** *Nothing drastic came out of Kyoto in the short-term, but the long-term goals are indeed coming to fruition now.*

2005-PRESENT
THE <u>LIE</u> OF "SCIENTIFIC CONSENSUS"

With the entirety of the Democrat Party now behind the hoax and eager to impose new regulations and taxes to fight GW/CC, the propaganda machine ran hard with the myth of near-universal scientific consensus on GW/CC. In 2013, a highly publicized FAKE study was published which claimed that 97% of scientists were in agreement with the "settled science."

One by one, the really "smart" people and celebrity scientists puffed-up and made rich and famous by the media, lined up to lend their respected names to the great cause. There was **Bill Nye** the goofy Science Guy, affirmative-action "scientist" **Neil DeGrasse-Tyson,** the "talking" stiff in a wheelchair **Stephen Hawking, Bill Gates** of Microsoft, **Mark Zuckerberg** of Facebook, Nobel Prize winners, NASA scientists and other assorted personages from the tech and science world.

Well now, if so many "smart" people believe *(or <u>claim</u> to believe?)* in the "settled science" of GW/CC, then only an "uneducated" or a "crazy" person could deny the truth now, right? These high and mighty endorsements of the fake theory, combined with the media's childish belittling of disbelievers with the dismissive label, "Climate Change Denier," have combined to create a stifling academic, social and political environment in which even trained scientists, let alone us mere mortals, are now afraid to publicly voice even the slightest doubt about the dogma of GW/CC. This FEAR alone is evidence that the great "cause" of "saving the planet" is based on marketing and brainwashing, not science.

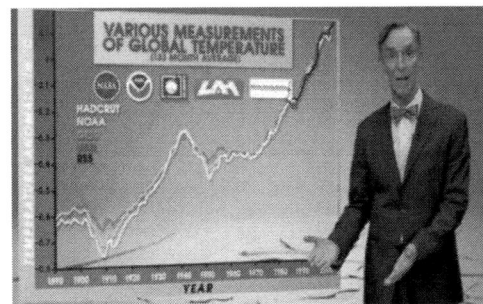

*The non-scientist **Bill Nye**, the grossly over-rated "astrophysicist" **Neil DeGrasse Tyson**, and Globalist **Bill Gates** want us to shut our minds down and just listen to them because they are supposedly smarter than us. Sadly, instead of using their own critical thinking faculties, many people do indeed base their own judgments on such opinions from "smart people."*

"The hand which rocks the cradle, rules the future."

In line with the stated objectives of **UNESCO -- The United Nations Educational, Scientific and Cultural Organization --** the entire educational establishment of the United States *(and Western Europe)* has joined the mighty movement to "save the planet." Just about every High School in America now has an "Environmental Science" curriculum.

No further words necessary. Just look at the images -- keeping in mind that this is only but a *fraction* of what's out there in our elementary schools and children's sections of bookstores / Amazon.

THIS IS PSYCHOLOGICAL CHILD ABUSE!

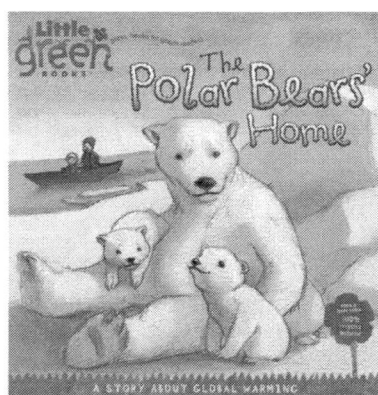

A large part of an entire generation -- many of them now adults - has been infected by this evil crap. By now, trying to dispel their deeply-held beliefs about "Global Warming" is like insulting their religious traditions!

CIRCA 2005
THE TERM "CLIMATE CHANGE" BECOMES
COMMONLY USED IN THE PUBLIC DISCOURSE

As the 20th Century came to a close, the **Remote Sensor Satellite** temperature readings **(RSS)** - which had shown a slight warming during the mid-1990's - no longer registered rising temperatures. At first, the pause in "Global Warming" was explained away as a temporary "anomaly," not in any way contradictory to the theory.

But as months of no warming turned into years of no warming, and even some very harsh winters, in America and Europe, the "scientists" adapted the "Global Warming" theory by arguing that extreme cold can also be a manifestation of the "crisis." And thus, the catch-all term "Climate Change" was more commonly used as a supplement, though not necessarily a replacement, for "Global Warming" -- a term which itself had replaced the silly 1980's slogan about some "Greenhouse Effect."

You see, actual greenhouses lock in heat mainly because their solid glass roofs magnify the sunlight while trapping heat with a *physical barrier*. The environmental analogy to greenhouses had to be replaced with the term "Global Warming" because "Greenhouse Effect" was just too quaint and idiotic.

"Climate Change" covers everything. Extreme cold, extreme heat, extreme storms, extreme rain, extreme droughts --- blame it on "Climate Change." Normal cold, normal heat, normal storms and normal rainfalls --- explain it all away as having to do with weather, not long-term climate. No matter what happens, "Climate Change" can be used to explain it.

1. A 1996 issue of Newsweek magazine assures its readers that recent "blizzards, floods and hurricanes" were due to "Global Warming." Years later, the term "Climate Change" was hatched by the marketing men behind this scam. 2. Even tornadoes are blamed on GW / CC now.

From "Greenhouse Effect" to "Global Warming" to "Climate Change" --- Make up your damn mind! In order to keep the scam afloat, the planners had to keep changing the name of the crisis to counter the contradictory data.

2006
AL GORE'S "AN INCONVENIENT TRUTH" BECOMES A BEST-SELLING BOOK AND MOVIE

In the closest Presidential Election race in history, Vice President **Al Gore** barely lost to George Bush back in 2000. The insanely ambitious Gore was so dejected that he grew a beard and lapsed into a severe state of depression while packing on about 50 pounds in one year. My goodness, Albert! Does fame and power really mean that much to you?

Gore finally regained his star status by publishing *An Inconvenient Truth* in 2006. Dealing solely with the "crisis" of what was by now referred to only as "Global Warming" *(the "Greenhouse Effect" label was no longer effective because it was too ridiculous-sounding)* the error-filled book *(more on that later)* also served as the basis for a slide-show documentary film version directed by **Davis Guggenheim** and narrated by Gore.

Gore's fake film premiered at the 2006 Sundance Film Festival. Both the book and the film were dutifully mega-hyped by the Piranha Press, with the book being

33

pushed to #1 on the "prestigious" New York Times best-selling list; and the documentary film winning two Academy Awards. Adding insult to idiocy, money-hungry / fame-hungry Gore was then awarded the Nobel Peace Prize in 2007 for his asinine activism.

*The world gone mad! Greasy Gore makes #1 on NY Times list, collects an Academy Award **and** a Nobel Peace Prize.*

2007
AL GORE'S BOOK IS CONVERTED INTO A CHILDREN'S VERSION AND TAUGHT TO CAPTIVE STUDENTS NATIONWIDE

Scholastic Books distributes books to schools nationwide. In 2007, the huge company really helped to push the "Global Warming" bandwagon in a major way. From the website of **Scholastic.com** (2007):

An Inconvenient Truth for Kids: Best-seller now available in kid-friendly version - By Aaron Broder *(kid reporter)*

"The ice caps at the North and South poles are melting, causing a gradual increase in the sea level. Last summer was one of the hottest on record. The number of hurricanes and tornadoes has been gradually increasing. What is the cause of all this unusual global activity? Two words: global warming.

Global warming is the subject of a new book for kids based on a best-seller for adults. An Inconvenient Truth has been adapted for kids ages 8–12 (Viking Children's, April 10, 2007, 92 pages)." **(5)**

By now, there was no escaping the constant wave of sophisticated propaganda being fed to the public, especially the vulnerable youth. The scam had gone "mainstream" -- with more and more skeptics keeping their doubts to themselves, for fear of being called "uneducated."

Young Aaron Broder, like so many other kids, was infected by the propaganda and passed on the virus to others.

* When your author here was that age, his "science teacher" was teaching about the coming ice-age!

The Day the Earth Stood Still is an adaptation of the 1951 science fiction film of the same name. This time around, the theme of the film was not the threat of nuclear war, but rather of "saving the planet" from man-made activities. Though the exact term "Global Warming" was never used, that was the common presumption of the audience. The film centered on the character "Klaatu" -- an alien sent to persuade humanity to reform its behavior toward the environment or face eradication at the mechanical hands of his giant robot.

The film, which cast A-list actor **Keanu Reeves** as Klaatu, benefitted from lots of hype and free publicity. During its opening weekend, and despite poor response from critics, it reached the #1 spot, grossing $30 million from 3,560 theaters. In spite of poor reviews, Reeve's box-office appeal and the initial media hype kept it in the top 10 for its first four weeks in theaters. The high and mighty planners behind the "Global Warming" scare understand very well the power of film as a medium for manipulating the public mind. And they have been "in cahoots" with the high and mighty of Leftist Hollywood for a long time.

Klatu will kill you if you don't stop driving that SUV.

Many millions of viewers and readers were exposed to the "We Can Solve It!" ad campaign. The corny ads featured political adversaries "who don't always agree," sitting on a love seat together and agreeing that the "crisis" of GW/CC had to be "solved." Liberal Democrats had been aboard the GW/CC Express since the scam's inception. The purpose of this propaganda campaign was to target Republicans and conservatives with the false notion that GW/CC wasn't a political movement, but rather, something that was in all of our interests to "solve."

Here is the dialogue from the ad which featured Democrat House Speaker *(at the time)* **Nancy Pelosi** and former Republican House Speaker, **Newt Gingrich**:

Pelosi: Hi. I'm Nancy Pelosi, lifelong Democrat and Speaker of the House.

Gingrich: And I'm Newt Gingrich, lifelong Republican and I used to be Speaker.

Pelosi: We don't always see eye to eye. Do we Newt?

Gingrich: No. But we do agree, our country must take action to address Climate Change.

Pelosi: We need cleaner forms of energy and we need them fast.

Gingrich: If enough of us demand action from our leaders, we can spark the innovation we need.

Pelosi: Go to wecansolveit.org. Together, we can do this. **(6)**

Another TV ad featured "conservative" reverend **Pat Robertson** and "liberal" reverend **Al Sharpton** on a beachfront loveseat, agreeing on the need to address GW/CC. And still another featured a female Democrat art student and a male Republican medical student, also agreeing, in spite of their different backgrounds, that GW/CC had to be "solved."

Cheesy and manipulative ads aimed at convincing people that the issue of GW/CC is not political.

1: *Pelosi & Gingrich*   **2.** *Sharpton and Robertson*

2008 – 2016
THE ELECTION AND PRESIDENCY OF BARACK OBAMA

The election of Senator **Barack Obama** (D-IL) -- henceforth to be un-affectionately referred to as "Obongo" -- to the US Presidency was a major advance for the "cause" of stopping non-existent GW/CC After nearly a quarter-century of gradual molding of the public mind, the hyped-up con-man didn't really have to sell the scam because it had already been accepted, to one degree or another, by slightly more than half of all voters. During a summer 2008 Berlin speech before a mob of 200,000 star-struck German imbeciles, the candidate spoke of "Global Warming" and "rising sea-levels" as if was already an established fact and declared his intentions to stop it. Get a load of this platitudinous puke:

As we speak, cars in Boston and factories in Beijing are melting the ice caps in the Arctic, shrinking coastlines in the Atlantic, and bringing drought to farms from Kansas to Kenya.

"Tonight, I speak before you not as a candidate for President, but as a citizen — a proud citizen of the United States, and a fellow citizen of the world....

This is the moment when we must come together to save this planet. Let us resolve that we will not leave our children a world where the oceans rise and famine spreads and terrible storms devastate our lands. Let us resolve that all nations –

including my own – will act with the same seriousness of purpose as has your nation, and reduce the carbon we send into our atmosphere. People of Berlin – people of the world – this is our moment. This is our time. **(7)**

In an interview with the San Francisco Chronicle, candidate Obongo openly stated that he intended to bankrupt coal companies and drive up electricity costs. An excerpt:

"I think clean air is critical and global warming is critical. ... (Under his plan) if somebody wants to build a coal-fired power plant, they can. It's just that it will bankrupt them because they are going to be charged a huge sum for all that greenhouse gas that's being emitted." Obama said, responding to a question about his cap-and-trade plan. **(8)**

He added:

"Under my plan ... electricity rates would necessarily skyrocket." **(9)**

Incredibly, under Obongo's EPA, CO_2 *(plant food)* was classified as a "pollutant" -- with the media cheering his destructive economic actions all the way. Private industry was thus subjected unprecedented and expensive regulations designed to fight a fictitious problem. Fortunately, and much to the delight of unemployed coal workers, President Trump, who had once "tweeted" that he believed Global Warming was a hoax, would later repeal some of these restrictions in 2017.

*1. Berlin: Obama tells 200,000 screaming imbeciles that he will lower sea-levels. **2.** Smug and arrogant Obongo coldly speaks of bankrupting companies and killing jobs without so much as batting an eyelash **3.** Trump knows it is all bullshit.*

Overrated clowns take a "selfie." -- **Barack Obama** *with Fake Scientists*
Bill Nye *and* **Neil DeGrasse Tyson**

MARCH, 2009
PRINCE CHARLES WARNS THAT THERE ARE ONLY 100 MONTHS LEFT TO "SAVE THE PLANET."

The world press announced the event with great fanfare when that elephant-eared worthless eater known as "The Prince of Wales" added his name to the all-star cast of high & mighty promoting GW/CC mythology.

The Herald of Scotland (*March 12, 2009*): Prince Charles: 100 months to avert catastrophic climate change

"Prince Charles warned last night that mankind has eight years or less to save the planet from a climate-created disaster.

*The heir to the throne told 150 business leaders in Rio de Janeiro, Brazil, that "the best projections tell us that we have **less than 100 months** to alter our behavior before we risk catastrophic climate change and the unimaginable horrors that this would bring". (emphasis added)*

He added: "Any difficulties which the world faces today will be nothing compared to the full effects global warming will have on the worldwide economy. It will result in vast movements of people escaping either flooding or droughts, in uncertain production of foods and lack of water and, of course, increasing social instability and potential conflict.

"It will affect the wellbeing of every man, woman and child on our planet." **(10)**

That 100 month warning bumped up the new doomsday to July, 2017 – which came and went. Fortunately for Charles, in 2015, he had updated his "100 months" prediction to "in another 30 years."

Daily Mirror *(July 19, 2015)*: Prince Charles Says We Have Just 30 Years to Save the Planet from Catastrophe (11)

Prince Charlie 2009: Only 8 years left to "save the planet."
Prince Charlie 2015: Only 30 years left to "save the planet."

2009: NASA warmist James Hansen: "four years to save the world." **(12)**

DECEMBER, 2009
THE COPENHAGEN CLIMATE CONFERENCE

The **2009 United Nations Climate Change Conference** was held in Copenhagen, Denmark. A general framework document for climate change mitigation beyond 2012 was to be agreed there. It recognized GW/CC as a crisis and stated that actions should be taken to keep temperature increases to a minimum. The ridiculous document was not legally binding and did not contain any commitments for reducing CO_2 emissions. Nonetheless, the hyped-up conference --- dubbed "Hopenhagen" by warmists, did provide yet another propaganda boost to the GW/CC movement.

On the final day of "Hopenhagen," America's buffoonish rookie President -- who ran on a platform of "hope" and of stopping the "rise of sea levels" -- addressed the assembled crooks and fools. More platitudinous puke:

"Good morning. It's an honor to for me to join this distinguished group of leaders from nations around the world. We come together here in Copenhagen because climate change poses a grave and growing danger to our people. You would not be here unless you – like me – were convinced that this danger is real. This is not fiction, this is science. Unchecked, climate change will pose unacceptable risks to our security, our economies, and our planet. That much we know.

So the question before us is no longer the nature of the challenge – the question is our capacity to meet it. For while the reality of climate change is not in doubt, our ability to take collective action hangs in the balance."

I believe that we can act boldly, and decisively, in the face of this common threat. And that is why I have come here today. As the world's largest economy and the world's second largest emitter, America bears our share of responsibility in addressing climate change, and we intend to meet that responsibility." **(13)**

Obongo leads the pack of self-important Globalist ass-clowns at "Hopenhagen."

2015
THE CATHOLIC POPE JUMPS ON THE "CLIMATE CHANGE" BANDWAGON

The Argentinean crypto-Marxist **Pope Francis** issued an unprecedented encyclical *(a softer form of an edict)* titled ***Laudato Si***, which essentially directed his 5,000 priests to accept that the "sin" of "Climate Change" is real and they should instruct their church members to join the call for "action." Though not nearly as influential of a figure as in decades or centuries past, the Catholic Pope still has a massive international following and carries a high degree of "moral authority," even for non-Catholics.

The shockingly liberal Pope's unprecedented endorsement of what, at its core, is really a *political* matter, gave a huge boost to the push for the grand international "Climate Agreement" which was to come in Paris in 2016. In advance of Paris 2016, the contemptible Leftist activist who masquerades as a "holy man" shamelessly exploited the suffering of people in the storm-ravaged Philippines by underhandedly linking a deadly 2013 typhoon to "Climate Change." The fake Pope declared:

"It is man who continuously slaps down nature. We have in a sense taken over nature. I think we have exploited nature too much. I think man has gone too far ...

The meetings in Peru were nothing much, I was disappointed. There was a lack of courage. We hope that in Paris the representatives have more courage to go forward." **(14)**

The Pope's support of the grand hoax gave it a *huge* propaganda boost and brought in many new believers – including this writer's devoutly Catholic 85-year-old Italian aunt who now fancies herself a climate expert. Mamma mia! This is madness!

Image 1: The Globalist Pope with a Leftist political history (Liberation Theology) hinted that the 2013 typhoon which devastated The Philippines was due to "carbon emissions."

"It is urgent to develop policy so that in the coming years, we drastically reduce carbon dioxide and other highly polluting gas emissions." **(15)**
-- Pope Francis

DECEMBER, 2015
THE PARIS CLIMATE AGREEMENT

The **Paris Climate Agreement** *(aka* ***Paris Climate Accord)*** is an agreement within the United Nations pre-existing framework. Parties to the deal are bound to mitigate greenhouse gas emissions and pursue "green energy" alternatives starting in the year 2020. As of June 2017, 150 states have ratified it.

The agreement requires that each country regularly report on its own contribution to mitigate "Global Warming." Though there is no international enforcement mechanism to compel compliance, politicians in countries across the world now have a legal basis for adapting policies within their respective nations -- expensive restrictions which can indeed be imposed by government force, albeit on a national level. This is what makes the Paris scheme so dangerous.

Speaking with one voice, the Globalist media is the US and the EU showered the big event with a TON of publicity.

Because the US Constitution requires a 2/3 Senate majority in order to approve a treaty, President Obongo had to issue an illegal "Executive Order" to circumvent the law and trap America into the Paris Agreement. The dwindling minority of Americans who still hold any respect for the Constitution were horrified at Obongo's power-grab. But the propaganda ministers at the *New York Times* and the rest of the Piranha Press were quick to praise him for his "leadership" in fighting the "crisis." The actual signing of the Paris Agreement by the parties was not until April 22nd, 2016 **(Earth Day).** It took place at the United Nations.

*Earth Day 2016: US Secretary of State **John Kerry** shamelessly used his granddaughter as a prop at the UN signing ceremony.*

In 2017, President Trump, who once correctly referred to "Global Warming" as a "hoax" but has since softened his position for political reasons, withdrew the United States from the agreement. This caused widespread condemnation among Globalists in the US and Europe as well as from Obongo and Pope Francis. In spite of Trump's withdrawal, many Democrat Mayors and Governors across America announced their intention to use the power of their offices to put the costly climate goals of Paris 2016 in effect in their jurisdictions. And it's quite possible that a future US President can rejoin. The Paris scam is therefore still a dangerous scheme.

*1. April 22, 2017: A highly-publicized Earth Day **"March for Science"** was organized to attack Trump for pulling the US out of the Paris trap. 2. The following week, the highly-publicized **"People's Climate March"** (led by **Leonardo DiCaprio (Image 3))** again attacked Trump over Paris. Who is really behind these players and their "spontaneous" events, and why? We'll get to that in Section 5!*

2016
LEONARDO DICAPRIO MAKES A NATIONAL GEOGRAPHIC DOCUMENTARY ABOUT "CLIMATE CHANGE"

Hollywood super-star **Leonardo DiCaprio**'s "crockumentary" about "Climate Change" marks his introduction as the new face of the movement to "save the planet." The pack of lies, massively hyped by the Piranha Press, is titled, ***Before the Flood***.

The film premiered at the Toronto International Film Festival in September 2016 before airing on the **National Geographic Channel** a few weeks later. To increase viewership, DiCaprio's propaganda was made widely available and free-of-charge on several different platforms.

At the European premiere in London in October 2016, DiCaprio introduced the film as follows:

"We went to every corner of the globe to document the devastating impacts of climate change and questioned humanity's ability to reverse what may be the most catastrophic problem mankind has ever faced. There was a lot to take on. All that we witnessed on this journey shows us that our world's climate is incredibly interconnected and that it is at urgent breaking point." **(16)**

Lies! And how much of this "dangerous" CO_2 do you and your crew emit by flying to *"every corner of the globe,"* eh Professor DiCaprio?

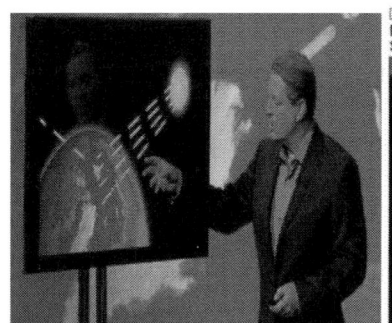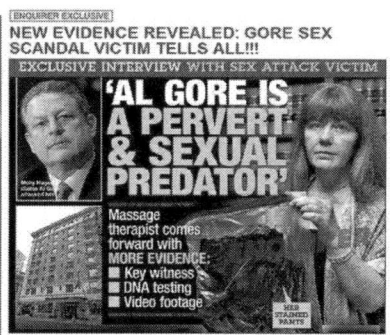

A few years after his embarrassingly perverted sexual assault scandal -- in which he was accused of forcefully masturbating on a professional masseuse -- and his wife's subsequent divorcing of him, Gore's role as the main public face of the hoax was assumed by the younger DiCaprio -- shown with co-conspirator Pope Frankie the Fake in Image 3.

SECTION 2

THE FAKE SCIENCE OF GLOBAL WARMING / CLIMATE CHANGE

"If anyone says, 'it (Climate Change) is just a theory,' it means they're missing a piece of their education where they do not fully understand what science is and how and why it works" (1)

- Neil deGrasse Tyson / Made for TV Affirmative Action pretend-scientist

COMMENTARY

Even the most dishonest warmist would have to concede that, in terms of public opinion, Fake News, Hollywood, and cartoonish propaganda for school children have always been the main drivers of "public opinion" regarding this issue. As former US Senator **Timothy Wirth** (D-CO) said in 1993:

 *"We've got to ride the global warming issue. **Even if the theory of global warming is wrong**, we will be doing the right thing, in terms of economic policy and environmental policy."* **(2)**

You see; the proverbial "they" -- unlike the true believers who have been brainwashed by Fake News, movies, cartoons and ignorant High School teachers – all know that the "science" of GW/CC is completely fake. But *you* aren't supposed to know it!

Apart from the core group of politically motivated or just simply bought-and-paid for scientists-for-hire *(many of whom indeed brilliantly clever)*, there is another class of self-styled "scientists" who, lacking in philosophical education and capable of little else but mimicking someone else's math equations, actually *do* believe in the fairy tale of GW/CC. These conceited clowns with Phds. can be even more fanatical and mentally unhinged the central pranksters – particularly if you undress them with basic facts.

The true agenda of GW/CC must remain hidden, as best as possible, at all times; which is why Senator Wirth, in a rare moment of honesty among friends, said what he did. He knows the science is fake. It's time you know it too.

50

INCONVENIENT TRUTH #1
THE "SCIENCE" OF MANMADE GW/CC IS THEORETICAL, NOT EXPERIMENT-BASED.

Not all forms of "science" are created equal. Before we even begin to unmask the major lies and fallacies of GW/CC, it is important to explain the difference between experimental science and theoretical science. Experimental science is traditional science which meets the rigorous standards of what has been revered for centuries as the **"Scientific Method."** It consists of three main elements: 1) Hypothesis 2) Experiment *(to test the hypothesis)* 3) Intense observation of the results. --- HEO -- Hypothesis, Experimentation, Observation --- always followed by repetition and peer review. Got it?

Theoretical science, on the other hand, consists mainly of math equations and computer models based on "proxy data." When utilized correctly, the various calculations can be useful for eliminating certain possibilities by deduction, or helping to craft a plausible hypothesis. But when pushed beyond its capacity and used for the purpose of artificially circumventing the Scientific Method, a scientist, drunk on his mathematical prowess, may soon drift down a path of folly and even insanity. Mathematics is only the language of science, but it is not science in and of itself unless **experimentation** and **observation** can support it.

A case in point: The early aviators used mathematics to design their primitive planes and calculate variables such as lift and drag *(theoretical science)*. Though they no doubt solved their formulas accurately, when they tried to fly their contraptions off the top of a hill, they often crashed after only a few yards of flight *(experimental science)*. The math was good, but the formula was wrong. And so, it was "back to the drawing board," aka "back to formula."

Nikola Tesla -- who, in this writer's judgment was, hands down, the greatest scientific genius and inventor of the 20th century *(Google him!)* -- warned about what he saw as a disturbing trend among modern scientists to substitute theoretical science for the Scientific Method. Tesla cautioned:

*"Today's scientists have substituted mathematics for experiments, and they wander off through equation after equation, **and eventually build a structure which has no relation to reality**.....*

These scientists think deeply, but not clearly. 'One must be sane to think clearly, but one can think deeply and be quite insane." **(3)**

Tesla spoke those words nearly 80 years ago, and the problem has grown far worse since. Entire generations of science students, though quite brilliant in mathematics, have been philosophically mis-trained in this respect. One can formulate and solve the most elaborate, dazzling and complex equations imaginable, and program a computer to forecast future phenomenon, but if the underlying formula has no connection to reality, it's all just an exercise in mathematical masturbation. And that's all that the "science" of GW/CC has going for it is concocted equations and rigged computer models to "predict" our flooded-out future. The CO_2-centric models miss so many other more significant climate variables that they have no use whatsoever.

Proxy Data

The dubious data which the number-crunching warmists feed into their fraudulent formulas is openly referred to as **"proxy data."** As it is with math equations, certain forms of proxy data can have limited uses, but it too has its limitations. Because the thermometer as well as the sophisticated instruments for measuring CO_2 weren't around back in the old days, warmists use proxy data found in tree rings, ice core samples glaciers and ice sheets, growth layers in coral, and sediment layers from the bottoms of lakes and oceans to estimate what the environment was like.

From this misused proxy data, the theoretical warmist magically concocts elaborate math equations, plugs his "finding" it into a rigged "computer model," hits the "enter" key and *voila!* When the computer "proves" that New York City and Washington DC will be under water if manmade emissions of carbon dioxide *(plant food)* are not controlled by government action, the well-funded warmists of academia start shouting: *"Science! Science!"* Express even the slightest doubt as to the viability of their "proxy data" driven video games and they *(and the media)* all start screaming, like little bitches: *"Climate Change Denier! "Climate Change Denier!"*

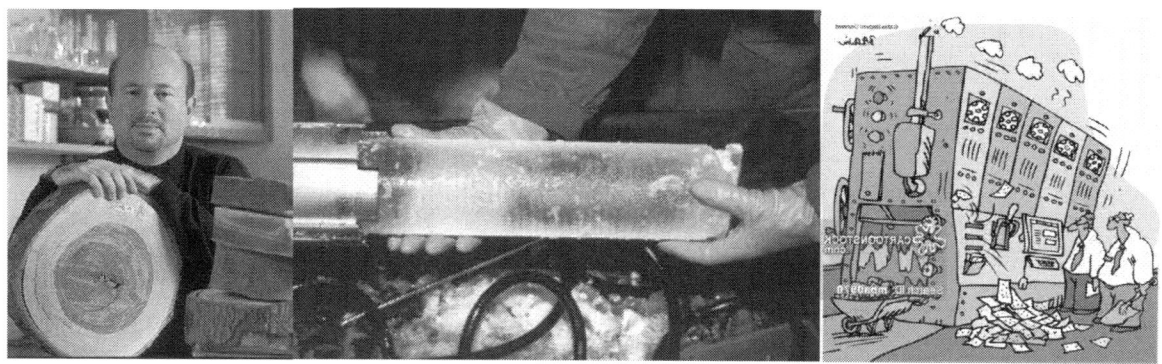

GARBAGE IN -- GARBAGE OUT

1. Warmist **Michael Mann** of Penn State shows off his tree rings which act as a historical "proxy" for the missing thermometer readings of years gone by. **2.** Ice Core samples to calculate CO_2 from year gone by. **3.** The incomplete and manipulated data all gets fed into the magic-math-machine and out pops "science."

No amount of math-magician tricks can substitute for *real* experimental science which uses *real* data. In fact, the GW/CC itself doesn't even lend itself to experimentation. Apart from there being no natural precedent to study, how would we possibly even go about creating a perfect replica mini-planet to experiment with? So, the next time a dogmatic warmist starts shouting: *"Science! Science! Science!* in your face, take a deep breath and calmly ask him or her the following question:

"Can you cite for me the name and date of the actual physical experiment which conclusively proved that a few added parts-per-million of atmospheric CO_2 will cause Antarctica to melt and flood the world's coastal cities?"

If the warmist doesn't flip out on you or run away from you *(I have triggered both reactions from warmists!)* -- then calmly school him, no matter what level of "education" he boasts of, on the critical distinction between theoretical science and experimental science. And if the warmist boasts of some level of scientific education, and starts babbling about unintelligible physics that *"you just wouldn't understand,"* calmly ask him:

"I do not doubt your superior mathematical prowess, but what evidence do you have that the physics and computer modeling of GW/CC correspond to reality?"

At that point, you will have unmasked the self-important genius -- checkmate!

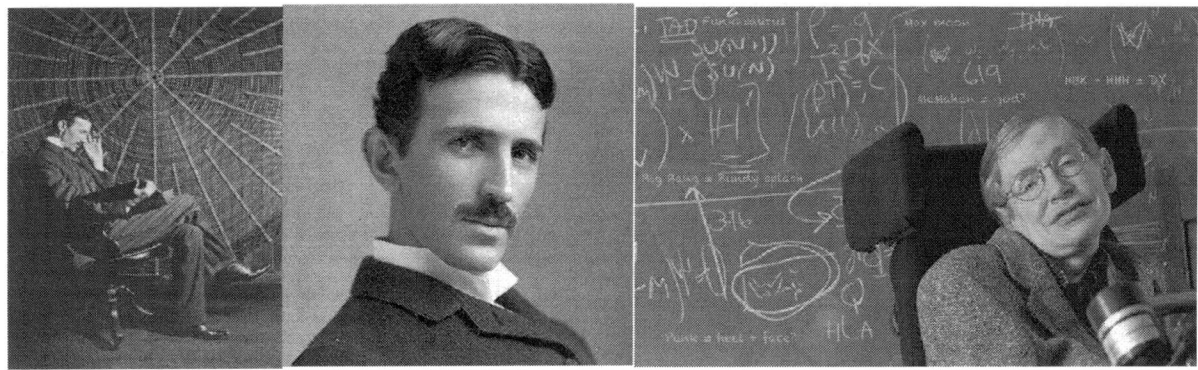

*Tesla, (Image 1 & 2) was an experimental scientist. He gave us commercial electricity and wireless communication. **Stephen Hawking** (the stiff in Image 2) is a theoretical scientist. He (or his keepers) gave us math equations about the unobservable "Big Bang" which prove nothing. See the difference?*

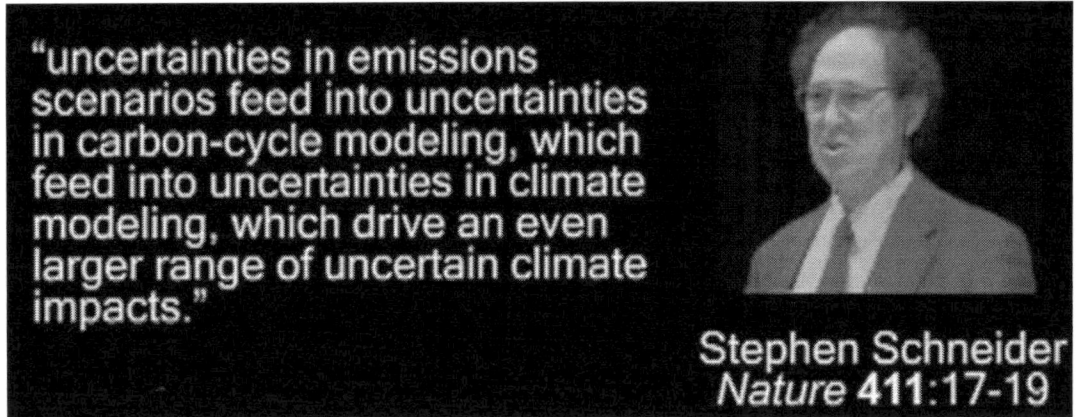

"uncertainties in emissions scenarios feed into uncertainties in carbon-cycle modeling, which feed into uncertainties in climate modeling, which drive an even larger range of uncertain climate impacts."

Stephen Schneider
Nature **411**:17-19

*Global Warmist **Stephen Schneider** puts his foot in his mouth and reveals to us the folly of theoretical computer modeling to predict climate trends.*

INCONVENIENT TRUTH #2
THERE HAD BEEN ZERO "GLOBAL WARMING" FOR TWENTY CONSECUTIVE YEARS UNTIL A STRANGE "ADJUSTMENT" WAS MADE

Satellite-based **Remote Sensor System (RSS)** have long been considered the "gold standard" for global temperature readings. Not only are readings immune from rigging, but the systems take so many measurements of so many different land and sea areas of the world and its atmosphere that is renders a true random sample which "old fashioned" land-based readings -- *which only cover a portion of the Earth's land surface* -- simply cannot even begin to match for accuracy.

During the mid-1990's, when RSS data revealed a very slight warming, the warmists breathlessly screamed, *"Aha! Science! Science! We told you this would happen!"* But after 1996, however, RSS data no longer showed that the slight warming had stopped. Since that time, mainly due to the rapid industrialization of countries such as China and India, manmade CO_2 emissions have increased by at least 25%. And yet, the zero-warming trend has continued, month after month, for nearly 20 straight years now.

The high priests of "Global Warming" never denied this data. How could they? They themselves often referred to the RSS flat temp data as "the pause." The way the warmists worked around the satellite numbers was to claim that the long term warming trend would still take place, and that "the pause" was just a statistical "anomaly." But when this "anomaly" extended to five consecutive years, then ten, then fifteen, they needed a new explanation as to why the planet was not warming.

Senator **Ted Cruz** (R-TX) brought national attention to "the pause" with his humiliating grilling of **Aaron Mair,** of the Sierra Club *(an environmentalist group).* Mair was powerless to explain away the then-18-year flat-lining of global temperatures. *(video available at Youtube, almost painful to watch.)*

1. In pressing Mair on "the pause," Cruz was able to tie the Leftist know-nothing into knots. 2. Mair with his friend Obama

For a while, the warmists tried to float a theory that the "missing heat" was being absorbed into the oceans.

Yale University *(March 30, 2015)*: **How Long Can Oceans Continue To Absorb Earth's Excess Heat? (4)**

Another variation was the claim, based on a "study" that CO_2 itself was being absorbed by the oceans, causing them to "acidify" and cause catastrophic harm oceanic ecosystems and wildlife.

New York Times *(January 15, 2015)*: **Ocean Life Faces Mass Extinction, Broad Study Says (5)**

An Australian warmist had announced the results of his "study" which explained away "the pause" as the result of unusually strong winds in the Pacific Ocean.

ABC News (Australia): *(February 9, 2014)*: Global warming: Australian scientists say strong winds in Pacific behind pause in rising temperatures (6)

But the newly concocted fairy tales would be difficult to sell, and marked a radical departure from the melting polar ice narrative which they had already invested so much time, money and energy in. What's a warmist propagandist to do about this inconvenient two-decade "pause?"

The warmists decided to simply ignore the RSS data and rely exclusively on land-based readings *(which can easily be falsified by cherry-picking locations or adjusting devices)*. But nagging questions about the discrepancy between RSS and land-based data simply would not go away. Oh those annoying "climate deniers" and their "obsession" with the RSS "pause!"

Finally, in March, 2016 -- **just one month prior to the signing of the Paris Agreement at the UN** -- the warmists made the audacious move to announce to the scientific community that the satellite readings had been faulty all along and needed to be "adjusted!" And thanks to the complicit media, they got away with making this outrageous "adjustment."

On March 7, 2016, A headline in the **Washington Post** mockingly blared:

Ted Cruz's Favorite Argument about Climate Change Just Got Weaker (7)

Phys.org gives a more detailed explanation:

(March 14, 2016): **Revamped Satellite Data Shows No Pause in Global Warming**

"The Remote Sensing System temperature data, promoted by many who reject mainstream climate science, now shows a slight warming of about 0.18 degrees Fahrenheit since 1998. Ground temperature measurements, which many scientists call more accurate, all show warming in the past 18 years.

"There are people that like to claim there was no warming; they really can't claim that anymore," said Carl Mears, the scientist who runs the Remote Sensing System temperature data tracking.

*The change resulted from an **adjustment** Mears made to fix a nagging discrepancy in the data from 15 satellites. The satellites are in a polar orbit, so they are supposed to go over the same place at about the same time as they circle from North to South Pole. Some of the satellites drift a bit, which changes their afternoon and evening measurements ever so slightly. Some satellites had drift that made temperatures warmer, others cooler. Three satellites had thrusters and they stayed in the proper orbit so they provided guidance for adjustments.*

*Mears said he was "motivated by fixing these differences between the satellites. If the differences hadn't been there, I wouldn't have done the upgrade." **(8)***

And just like that, an "adjustment" was made -- **we repeat, one month prior to the signing of the Paris Agreement** -- so that RSS data could show a *tiny* 0.18 °F increase since 1998. How convenient! Problem solved. Sorry, boys and girls. This advanced student of history, geopolitics and human nature ain't buyin' Professor Mears' oddly-timed "adjustment."

And notice how the "adjustment" was made so small in order to avoid the appearance of blatant rigging. Nonetheless, the goal of turning a zero into a positive, albeit an insignificant positive, was accomplished. But even if one, only for the sake of argument, were to accept this convenient 0.18 °F bump as legitimate; in essence, a zero-20-year-warming trend, confirmed by the warmists, remains in effect -- and counting.

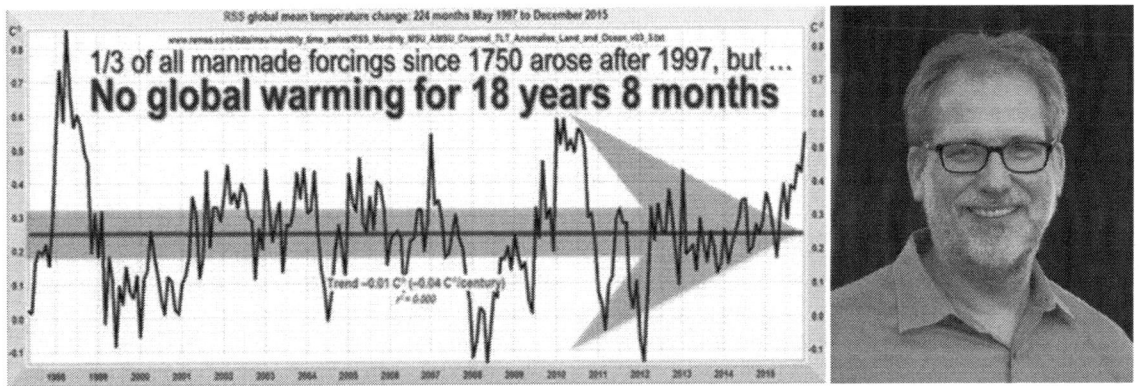

*The highly embarrassing hard data which showed **zero warming** for nearly 20 straight years could not be refuted. When alternative explanations failed to adequately explain away "the pause," Professor Mears simply "adjusted" the satellite readings to show a meaningless 0.18 °F degree "warming."*

INCONVENIENT TRUTH #3
ANTARCTIC SEA ICE HAS NOT CHANGED OVER THE PAST 100 YEARS

From the **London Telegraph** / **Science Section** / *November 24, 2016* :

Scott and Shackleton logbooks prove Antarctic sea ice is not shrinking 100 years after expeditions

Antarctic sea ice had barely changed from where it was 100 years ago, scientists have discovered, after poring over the logbooks of great polar explorers such as Robert Falcon Scott and Ernest Shackleton.

Experts were concerned that ice at the South Pole had declined significantly since the 1950s, which they feared was driven by man-made climate change.

But new analysis suggests that conditions are now virtually identical to when the Terra Nova and Endurance sailed to the continent in the early 1900s, indicating that declines are part of a natural cycle and not the result of global warming.

--- "The missions of Scott and Shackleton are remembered in history as heroic failures, yet the data collected by these and other explorers could profoundly change the way we view the ebb and flow of Antarctic sea ice," said Dr Jonathan Day, who led the study, which was published in the journal The Cryosphere.

"We know that sea ice in the Antarctic has increased slightly over the past 30 years, since satellite observations began. Scientists have been grappling to understand this trend in the context of global warming, but these new findings suggest it may not be anything new." **(9)**

And then there's this observation from **NASA's own website,** October 30, 2105 *(oops! somebody at NASA went off script):*

NASA Study: Mass Gains of Antarctic Ice Sheet Greater than Losses

"A new NASA study says that an increase in Antarctic snow accumulation that began 10,000 years ago is currently adding enough ice to the continent to outweigh the increased losses from its thinning glaciers.

The research challenges the conclusions of other studies, including the Intergovernmental Panel on Climate Change's (IPCC) 2013 report, which says that Antarctica is overall losing land ice." **(10)**

The warmists can spin these inconvenient facts, confirmed by their beloved NASA, no less, as much as they like --- but the fact of the matter is, **Antarctica is not melting away.** And that is why, in December of 2013, a ship full of well-paid warmist "researchers" got stuck in the ice and needed to be rescued!

Washington Times: *(January 5, 2014)* **Irony alert: Global Warmists get stuck in ice (11)**

So, what is a creative warmist to do to explain away this bit of inconvenient data? Check out this slick maneuver:

From the Daily Mail (UK) / *(April 1, 2013)***:**

*"Climate change experts have been trying for years to explain why the sea ice in Antarctica is expanding. Now scientists claim to have found the answer – global warming. They believe the **paradoxical shift** is caused by water melting from beneath the Antarctic ice shelves and refreezing back on the surface."* **(12)**

In other words, melting ice caps cause expanding ice caps. Who knew? Shrinking ice caps --- blame "Global Warming." Expanding ice caps --- blame "Global Warming." The climate con-men have got an answer for everything.

*1 & 2. The detailed log books of Captains **Ernest Shackleton** and **Robert Scott** match up very closely with current NASA satellite observations. 3. No substantive change in the floating ice-shelf for 100 + years! 4. Warmist researchers stuck and surrounded by Antarctic ice.*

> ## INCONVENIENT TRUTH #4
> ## ARCTIC SEA ICE CAN VARY GREATLY FROM YEAR-TO-YEAR AND THERE HAS BEEN NO NOTICEABLE PERMANENT CHANGE OVER THE PAST 80 YEARS

Whereas the South Pole is land buried beneath an average of 2000 meters of ice and surrounded by water, the North Pole is basically water, often covered with just a few meters of ice, and surrounded by land. Greenland, much like Antarctica, is land covered by deep ice, except in certain areas which are actually inhabited.

The warmists, who are known for cherry-picking data, are quick to pounce on Arctic temperature and ice-coverage data from years in which there is diminishing ice, but they conveniently ignore other years in which the ice coverage is thicker and wider. By the way, ice that sits on top of water would not cause rising sea-levels anyway – just the opposite.

Simple Experiment: Fill a glass with ice cubes, then fill it with water - right up to the lip. Then, place this in a warm place and wait for the ice to melt. What do you have after the ice melts? LESS volume in the glass than before, as ice is expanded when frozen. Therefore, a melting ice-shelf *(which sits atop water)* would *lower* the level of the sea, not raise it!

Of course, this whole discussion is a mute point because we have already established that global temperatures are not warming. Nonetheless, NASA, though it continues to promote the "Global Warming" hoax, contradicts itself with hard data.

Forbes Magazine *(May 19, 2015)*: **Updated NASA Data: Global Warming Not Causing Any Polar Ice Retreat**

"Updated data from NASA satellite instruments reveal the Earth's polar ice caps have not receded at all since the satellite instruments began measuring the ice caps in 1979. Since the end of 2012, moreover, total polar ice extent has largely remained above the post-1979 average. The updated data contradict one of the most frequently asserted global warming claims – that global warming is causing the polar ice caps to recede.

During the modest decline in 2005 through 2012, the media presented a daily barrage of melting ice cap stories. Since the ice caps rebounded – and then some – how have the media reported the issue?" **(13)**

The article inked to a chart showing the NASA data. There were some ups and downs, but nothing terrible above or below the baseline averages of the past 50 years.

As compelling as the hard data may be, what really obliterates the lie of Arctic warming into a million pieces are news stories from many years ago -- reported during a time -- 1922 to be precise -- when man-made CO_2 emissions was but a tiny fraction of what is released today. It certainly doesn't take a "scientist" to appreciate the significance of past recorded events in the Arctic.

Arctic Ocean Getting Warm; Seals Vanish And Icebergs Melt

(By the Associated Press.)

The Arctic ocean is warming up, icebergs are growing scarcer and in some places the seals are finding the waters too hot, according to a report to the Commerce Department yesterday from Consul Ifft, at Bergen, Norway.

Reports from fishermen, seal hunters and explorers, he declared, all point to a radical change in climatic conditions and hitherto unheard-of temperatures in the Arctic zone. Exploration expeditions report that scarcely any ice has been met with as far north as 81 degrees 29 minutes. Soundings to a depth of 3,100 meters showed the gulf stream still very warm.

Great masses of ice have been replaced by moraines of earth and stones, the report continued, while at many points well known glaciers have entirely disappeared. Very few seals and no white fish are being found in the eastern Arctic, while vast shoals of herring and smelts, which have never before ventured so far north, are being encountered in the old seal fishing grounds.

THE CHANGING ARCTIC.

By GEORGE NICOLAS IFFT.

[Under date of October 10, 1922, the American consul at Bergen, Norway, submitted the following report to the State Department, Washington, D. C.]

The Arctic seems to be warming up. Reports from fishermen, seal hunters, and explorers who sail the seas about Spitzbergen and the eastern Arctic, all point to a radical change in climatic conditions, and hitherto unheard-of high temperatures in that part of the earth's surface.

In August, 1922, the Norwegian Department of Commerce sent an expedition to Spitzbergen and Bear Island under the leadership of Dr. Adolf Hoel, lecturer on geology at the University of Christiania. Its purpose was to survey and chart the lands adjacent to the Norwegian mines on those islands, take soundings of the adjacent waters, and make other oceanographic investigations.

Dr. Hoel, who has just returned, reports the location of hitherto unknown coal deposits on the eastern shores of Advent Bay—deposits of vast extent and superior quality. This is regarded as of first importance, as so far most of the coal mined by the Norwegian companies on those islands has not been of the best quality.

[37] R. L. Holmes: Quart. Journ, Royal Meteorol. Soc., January, 1905.

MONTHLY WEATHER REVIEW.

NOVEMBER, 1922.

The oceanographic observations have, however, been even more interesting. Ice conditions were exceptional. In fact, so little ice has never before been noted. The expedition all but established a record, sailing as far north as 81° 29' in ice-free water. This is the farthest north ever reached with modern oceanographic apparatus.

The character of the waters of the great polar basin has heretofore been practically unknown. Dr. Hoel reports that he made a section of the Gulf Stream at 81° north latitude and took soundings to a depth of 3,100 meters. These show the Gulf Stream very warm, and it could be traced as a surface current till beyond the 81st parallel. The warmth of the waters makes it probable that the favorable ice conditions will continue for some time.

Later a section was taken of the Gulf Stream off Bear Island and off the Isfjord, as well as a section of the cold current that comes down along the west coast of Spitzbergen off the south cape.

In connection with Dr. Hoel's report, it is of interest to note the unusually warm summer in Arctic Norway and the observations of Capt. Martin Ingebrigtsen, who has sailed the eastern Arctic for 54 years past. He says that he first noted warmer conditions in 1918, that since that time it has steadily gotten warmer, and that to-day the Arctic of that region is not recognizable as the same region of 1868 to 1917.

Many old landmarks are so changed as to be unrecognizable. Where formerly great masses of ice were found, there are now often moraines, accumulations of earth and stones. At many points where glaciers formerly extended far into the sea they have entirely disappeared.

WARMER ARCTIC
Auckland Star, Volume LXXI, Issue 297, 14 December 1940

▸ About this newspaper   ▸ View computer-g

WARMER ARCTIC DISAPPEARING ICE. SCIENTISTS' REPORTS. THIRD LESS IN 50 YEARS.
(By THOMAS R. HENRY.)
WASHINGTON

The ice of the Arctic Ocean is melting so rapidly that more than one-third of it has disappeared in fifty years.

Anchorage Daily Times

MEMBER ASSOCIATED PRESS

VOL. VII. NO. 2.   ANCHORAGE, ALASKA, THURSDAY, NOVEMBER 2, 1922.   PRICE TEN CENTS

INDICATIONS ARCTIC MAY BECOME TEMPERATE ZONE

OBREGON'S ARCH ENEMY FALLS INTO HANDS OF FEDERAL FORCE NEAR CITY OF DURANGO, MEX.

GENERAL FRANCISCO MURGUIA PARTICIPATED IN BATTLE THAT BROUGHT DEATH TO FORMER PRESIDENT CARRANZA—END OF BANDIT'S CAREER CAME WHEN FEDERAL FORCE SURROUNDED MURGUIA AND FOLLOWERS.

(By Associated Press)

MEXICO CITY, Nov. 2.—General Francisco Murguia, arch enemy of President Obregon, and one of the men who took part in the fight that brought death to former President Carranza, has fallen into the hands of federal troops, and his career as a dangerous rebel is thought to be at an end, with a little band of followers which was surrounded yesterday near Durango City, it was announced at the president's office. Whether they will be taken to Durango City for trial or brought to the capital and publicly executed has not yet been determined.

HENRY FORD REPORTED NEGOTIATING FOR LARGE COAL REGION INVOLVING PRICE FIFTEEN MILLION DOLLARS

MORE THAN 180,000,000 TONS OF COAL CONSIDERED IN DEAL NOW PENDING.

(By Associated Press)

PITTSBURGH, Nov. 2.—More than eighty acres of coal land containing approximately 180,000,000 tons of bituminous coal is involved in a deal which is said to be in process of negotiation between Henry Ford and the Wayne Coal company. Coal men place the price in excess of $15,000,000.

PAY U. S. DEBTS SAYS CHANCELLOR

NEW BRITISH CABINET OFFICIAL STATES STAND ON MATTER OF WAR LOANS.

(By Associated Press)

AMBASSADOR PAGE IS REPORTED DEAD

FORMER MINISTER TO ITALY AND AUTHOR PASSES AWAY IN VIRGINIA.

(By Associated Press)

ALAMEDA SAILS.

SEATTLE, Wn., Nov. 2.—The S. S. Alameda sailed at 2 yesterday morning, carrying ninety eight passengers. Second passengers: Mrs. H. Dougherty and son, Mrs. A. H. Barker, Mrs. K. Hawkins, Capt. A. K. Lathrop, J. M. Ferguson, W. Breavy, George Nelson, W. Tervey, D. B. Card, Mrs. G. Shea, Elizabeth Taffinger, Mrs. Kate Taffinger, W. C. Kilborg, E. L. T. Rasvapouit, Christ Jennett and Leo Doyle.

UNUSUAL TEMPERATURE OF ARCTIC INDICATES REMARKABLE CHANGE TAKEN PLACE IN FROZEN ZONE

WATERS NORWEGIAN ARCTIC BECOMING TOO WARM FOR SEALS—GLACIERS DISAPPEARING—BEING REPLACED BY MORAINE EARTH AND STONES—HERRING AND SMELT VENTURE FARTHER INTO ARCTIC OCEAN.

(By Associated Press)

WASHINGTON, Nov. 2.—The Arctic ocean is warming up and icebergs are growing scarcer in some places, the seals finding the waters too hot, according to reports to the department of commerce from Consul Ift at Bergen, Norway. Reports from fishermen, seal hunters and explorers declare that at all points a radical change in climatic conditions is manifested. Unheard-of temperatures are reported in the Arctic zone. Exploration expeditions report scarcely any ice has been met as far north as 81 degrees, 29 minutes. Soundings to a depth of 3,100 meters

1922: "Seals Vanish" -- "Icebergs Melt" -- "Radical Change in Climate" -- "Disappearing Ice -- "Arctic Ocean Melting Rapidly" -- "Arctic May Become Temperate Zone -- "Remarkable Change"

So you see, boys and girls, not only is the Arctic not melting away at this time, but even it were, there is nothing to be afraid of. The seas-levels did not rise in 1922 and the ice eventually came back. So much so that in 2016 *another* team of warmists, who, evidently now believe their own bullshit about ice-free Arctic, also got stuck in the Arctic ice, just like their 2013 counterparts did in Antarctica.

The Daily Caller *(June 20, 2016)*: Global Warming Expedition Stopped In Its Tracks by Arctic Sea Ice

"A group of adventurers, sailors, pilots and climate scientists that recently started a journey around the North Pole in an effort to show the lack of ice, has been blocked from further travels by ice. The Polar Ocean Challenge is taking a two month journey .... Their objective, as laid out by their website, was to demonstrate "that the Arctic sea ice coverage shrinks back so far now in the summer months that sea that was permanently locked up now can allow passage through." **(14)**

Ya just can't make this stuff up!

Science & Mechanics November 1969

During the 1970's, "scientific" talk of a coming "ice age" *(often linked to pollution)* began to make the rounds of academia and media. Though it never reached the fever pitch of today's GW/CC, most of us over the age of 45 will remember this mini-scare concocted by "respected" scientists and reported by "respected" mainstream media.

Below is a list of articles and sources from that period. Many of the articles can be read in full by Googling the titles, but you may have to pay for archive access:

1970 – Colder Winters Herald Dawn of New Ice Age – Scientists See Ice Age In the Future (*The Washington Post, January 11*)

1970 – **Is Mankind Manufacturing a New Ice Age?** (*L.A. Times, January 15*)

1970 – New Ice Age May Descend On Man (*Sumter Daily Item, January 26, 1970*)

1970 – Pollution Prospect A Chilling One (*Owosso Argus-Press, January 26*)

The prospect is literally chilling. The ultimate in climate control — 20 degrees cooler not only inside but outdoors as well.

And if by now we are accustomed, if not inured, to the physical threat of pollution, along comes a warning there may also be dire political consequences.

Dr. Arnold Reitze, an expert in the legal aspects from Cleveland's Case Western Reserve University, suggests pollution, or the effort to control it, could be fatal to our concept of a free society.

As likely inevitable restraints on the individual and mass, Reitze suggests:

● Outlawing the internal combustion engine for vehicles and outlawing or strick controls over all forms of combustion.

● Rigid controls on the marketing of new products, which will be required to prove a minimum pollution potential.

● Controls on all research and development, to be halted at the slightest prospect of additional pollution.

● Possibly even population controls, the number of children per family prescribed and punishment for exceeding the limit.

In Reitze's view, "We will be forced to sacrifice democracy by the laws that will protect us from further pollution."

In article copied above, (January 26, 1970) a much younger Dr. Arnold Reitze, an "Environmental Law" professor, <u>openly called for a dictatorship</u> to control the CO2 emissions which he and others were saying would cause an ice-age. Reitze is a warmist now.

1970 – **Pollution's 2-way 'Freeze' On Society** (*Middlesboro Daily News, January 28, 1970*)

1970 – Cold Facts About Pollution (*The Southeast Missourian, January 29, 1970*)

1970 – **Pollution Could Cause Ice Age**, Agency Reports (*St. Petersburg Times, March 4, 1970*)

1970 – **Pollution Called Ice Age Threa**t (*St. Petersburg Times, June 26, 1970*)

1970 – **Dirt Will .Bring New Ice Age** (*The Sydney Morning Herald, October 19, 1970*)

1971 – Ice Age Refugee Dies Underground (*The Montreal Gazette, February 17, 1971*)

1971 – U.S. Scientist Sees New Ice Age Coming (*The Washington Post, July 9, 1971*)

1971 – Ice Age Around the Corner (*Chicago Tribune, July 10, 1971*)

1971 – New Ice Age Coming – It's Already Getting Colder (*L.A. Times, October 24, 1971*)

1971 – Another Ice Age? Pollution Blocking Sunlight (*The Day, November 1, 1971*)

1971 – **Air Pollution Could Bring An Ice Age** (*Harlan Daily Enterprise, November 4, 1971*)

1972 – Air pollution may cause ice age (*Free-Lance Star, February 3, 1972*)

1972 – Scientist Says New ice Age Coming (*The Ledger, February 13, 1972*)

1972 – Scientist predicts new ice age (*Free-Lance Star, September 11, 1972*)

1972 – British expert on Climate Change says Says New Ice Age Creeping Over Northern Hemisphere (*Lewiston Evening Journal, September 11, 1972*)

1972 – Climate Seen Cooling For Return Of Ice Age (*Portsmouth Times, September 11, 1972*)

1972 – New Ice Age Slipping Over North (*Press-Courier, September 11, 1972*)

1972 – Ice Age Begins A New Assault In North (*The Age, September 12, 1972*)

1972 – Weather To Get Colder (*Montreal Gazette, September 12, 1972*)

1972 – British climate expert predicts new Ice Age (*The Christian Science Monitor, September 23, 1972*)

1972 – Scientist Sees Chilling Signs of New Ice Age (*L.A. Times, September 24, 1972*)

1972 – Science: Another Ice Age? (*Time Magazine, November 13, 1972*)

1973 – The Ice Age Cometh (*The Saturday Review, March 24, 1973*)

1973 – Weather-watchers think another ice age may be on the way (*The Christian Science Monitor, December 11, 1973*)

1974 – New evidence indicates ice age here (*Eugene Register-Guard, May 29, 1974*)

1974 – Another Ice Age? (*Time Magazine, June 24, 1974*)

1974 – 2 Scientists Think 'Little' Ice Age Near (*The Hartford Courant, August 11, 1974*)

1974 – Ice Age, worse food crisis seen (*The Chicago Tribune, October 30, 1974*)

1974 – **Believes Pollution Could Bring On Ice Age** (*Ludington Daily News, December 4, 1974*)

1974 – **Pollution Could Spur Ice Age, NASA Says** (*Beaver County Times, December 4, 1974*)

1974 – **Air Pollution May Trigger Ice Age**, Scientists Feel (*The Telegraph, December 5, 1974*)

1974 – **More Air Pollution Could Trigger Ice Age Disaster** (*Daily Sentinel – December 5, 1974*)

1974 – **Scientists Fear Smog Could Cause Ice Age** (*Milwaukee Journal, December 5, 1974*)

1975 – Climate Changes Called Ominous (*The New York Times, January 19, 1975*)

1975 – Climate Change: Chilling Possibilities (*Science News, March 1, 1975*)

1975 – B-r-r-r: New Ice Age on way soon? (*The Chicago Tribune, March 2, 1975*)

1975 – Cooling Trends Arouse Fear That New Ice Age Coming (*Eugene Register-Guard, March 2, 1975*)

1975 – Is Another Ice Age Due? Arctic Ice Expands In Last Decade (*Youngstown Vindicator – March 2, 1975*)

1975 – Is Earth Headed For Another Ice Age? (*Reading Eagle, March 2, 1975*)

1975 – New Ice Age Dawning? Significant Shift In Climate Seen (*Times Daily, March 2, 1975*)

1975 – There's Troublesome Weather Ahead (*Tri City Herald, March 2, 1975*)

1975 – Is Earth Doomed To Live Through Another Ice Age? (*The Robesonian, March 3, 1975*)

1975 – The Ice Age cometh: the system that controls our climate (*The Chicago Tribune, April 13, 1975*)

1975 – Newsweek: The Cooling World (*Peter Gwynne, April 28, 1975*)

"There are ominous signs that the earth's weather patterns have begun to change dramatically and that these changes may portend a drastic decline in food production with serious political implications for every country on earth. The drop in food production could begin quite soon.

The central fact is that, after three quarters of a century of extraordinarily mild conditions, the Earth seems to be cooling down. Meteorologists disagree about the cause and extent of the cooling trend, as well as over its specific impact on local weather conditions. But they are almost unanimous in the view that the trend will reduce agricultural productivity for the rest of the century." **(15)**

1. Newsweek article from 1975 2 & 3. Peter Gwynne now regrets writing it.

1975 – Scientists Ask Why World Climate Is Changing; Major Cooling May Be Ahead (*The New York Times, May 21, 1975*)

1975 – In the Grip of a New Ice Age? (*International Wildlife, July-August, 1975*)
1975 – **Oil Spill Could Cause New Ice Age** (*Milwaukee Journal, December 11, 1975*)

1976 – The Cooling: Has the Next Ice Age Already Begun? [Book] (*Lowell Ponte, 1976*)
1977 – Blizzard – What Happens if it Doesn't Stop? [Book] (*George Stone, 1977*)
1977 – The Weather Conspiracy: The Coming of the New Ice Age [Book] (*The Impact Team, 1977*)
1976 – Worrisome CIA Report; Even U.S. Farms May be Hit by Cooling Trend (*U.S. News & World Report, May 31, 1976*)
1977 – The Big Freeze (*Time Magazine, January 31, 1977*)
1977 – We Will Freeze in the Dark (*Capital Cities Communications Documentary, Host: Nancy Dickerson, April 12, 1977*)
1978 – The New Ice Age [Book] (*Henry Gilfond, 1978*)
1978 – Little Ice Age: Severe winters and cool summers ahead (*Calgary Herald, January 10, 1978*)
1978 – Winters Will Get Colder, 'we're Entering Little Ice Age' (*Ellensburg Daily Record, January 10, 1978*)
1978 – Geologist Says Winters Getting Colder (*Middlesboro Daily News, January 16, 1978*)
1978 – It's Going To Get Colder (*Boca Raton News, January 17, 1978*)
1978 – Believe new ice age is coming (*The Bryan Times, March 31, 1978*)
1978 – An Ice Age Is Coming Weather Expert Fears (*Milwaukee Sentinel, November 17, 1978*)

1979 – Get Ready to Freeze (*Spokane Daily Chronicle, October 12, 1979*)
1979 – New Ice Age Almost Upon Us? (*The Christian Science Monitor, November 14, 1979*)

Some of the "scientists" even claimed that manmade Carbon Dioxide CO_2 -- today's cause of alleged GW/CC -- was the culprit behind the cooling world. The most visible report of this "coming ice age" was a 1978 episode of **In Search Of,** a popular TV show which explored mysterious phenomenon. The show was narrated by the oh-so-scary-sounding **Leonard Nimoy** *(Mr. Spock of Star Trek fame)*. Let me tell you, it darn near scared the crap out of this author, who was barely a teenager at the time.

Here is Nimoy's intro from the episode's trailer *(with eerie music and the constant sound of cold wind in the background)*:

"In 1977, the worst winter in a century struck the United States. Arctic cold gripped the Midwest for weeks on end. Great blizzards paralyzed cities of the northeast. One desperate night in Buffalo, eight people froze to death in marooned cars. Pat Bushnell was on the road that night."

(Ms. Bushnell speaking): Traffic just absolutely stopped. I was afraid of being stuck in the car all night long with the cold and the wind, running out of gas, and then what? I think that if we had to go through a real bad winter just like we just went through, we have to think about moving someplace else. **(16)**

Nimoy's serious voice resumes:

"Move where? The brutal Buffalo winter might become common all over the United States. Climate experts believe the next ice age is on its way. According to recent evidence, it could come sooner than anyone had expected.

At weather stations in the far north, temperatures have been dropping for thirty years. Sea coasts long free of summer ice are now blocked year round. According to some climatologists, within a lifetime, we might be living in the next ice age."

(Cue very scary music) **(17)**

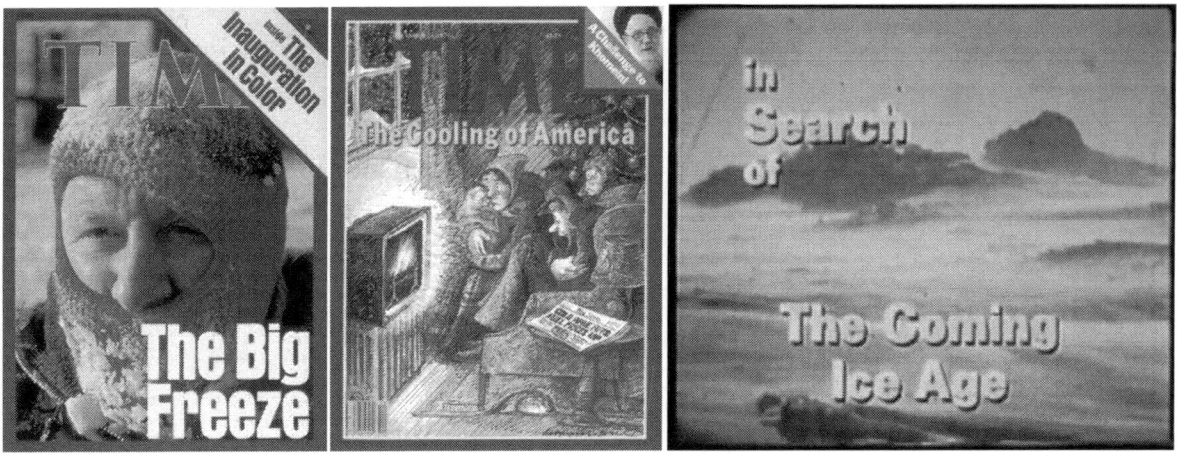

The 1970's ice-age scare got a fair amount of Establishment press, but was later abandoned for the Global Warming scenario.

.

Looking back on the 10-year mini-scare establishes an historical precedent which teaches us an important lesson -- namely, just because "experts" declare something to be so, it doesn't make it so. One of the "scientists" who was once on the ice-age bandwagon and then switched to promoting "The Greenhouse Effect / Global Warming" was the late **Stephen Schneider** of Stanford, *(one of Al Gore's gurus)*. In his case, the strange switch from a "Global Coolist" to a "Global Warmist" took place within a single year. In 1978, Schneider actually appeared on Nimoy's *In Search Of* show. In 1979, he was predicting that we could be launching boats from the steps of the US Capitol Building and that a third of the state of Florida would be under water! **(18)**

What a difference 1 year makes!

It is a matter, as Schneider explains, of statistics. The last 15,000 years have been unusually warm when compared to global temperatures for the last 150,000. The last 200 years have been unusually warm when compared to the last 1,000. But there is considerable evidence that this warm period is passing and that temperatures on the whole will get colder. For example, in the last 100 years mid-latitude air temperatures peaked at an all-time warm point in the 1940's and have been cooling ever since.

***Ice Age Schneider** in 1978 was an ice-ager, and appeared on TV as one.*

1. The same Schneider, by 1979, was a full-blown warmist. In 1990, his warmist book was puffed-up to the New York Times best-seller list.
2. Schneider in later years with his "student" Al Gore

INCONVENIENT TRUTH #7
CONTRARY TO THE CLAIMS OF THE MEDIA, THERE IS NO "SCIENTIFIC CONSENSUS" ON GW/CC

Ask any warmist why he believes in GW/CC and the he will most likely invoke the clichéd mantra about: *"The consensus of scientists."*

Apart from the fact that true science is based solely on hard facts and not some democratic "consensus," and apart from the fact that "consensus" is often just an indication of sheep-like "groupthink" or peer pressure *(very common amongst academic types),* the fact of the matter is that the claim of near-universal "consensus" is a bare-faced lie!

Richard Cohen, a widely read left-wing Democrat columnist for the Washington Post, offers us a perfect example of this type of manipulative intellectual bullying:

"There were some, of course, just as there are some scientists who are global warming skeptics, but these few- about 2% of climate researchers- could hold their annual meeting in a phone booth, if there are any left." **(19)**

There are two important contradicting items regarding this fictitious "consensus" that Cohen and his cadres of commie cohorts will not tell their gullible readers. You should know about them.

1) The Global Warming Petition Project

The **Global Warming Petition Project**, also known as the **Oregon Petition**, is a petition urging the United States government to reject policies aimed at stopping GW/CC because the signatories do not believe in man-made CO_2-based GW/CC. Since its establishment in 1998, nearly **31,500** scientists have signed the petition - 9,000 of them Phd's and many with backgrounds in atmospheric physics, climatology and meteorology. Physics legends **Edward Teller** *(deceased)* and **Freeman Dyson** are among the signatories of the petition which declared:

"There is no convincing evidence that human release of carbon dioxide, methane, or other greenhouse gases is causing or will in the foreseeable future cause catastrophic heating of the Earth's atmosphere and disruption of the Earth's climate. Moreover, there is substantial scientific evidence that increases in atmospheric carbon dioxide produce many beneficial effects upon the natural plant and animal environment of the Earth." **(20)**

Were it not for the relentless media and grant-money-driven pressure for university academics to get on board the GW/CC Express, we can only imagine how many more fearful scientists would come out of the closet to add their names to this massive list of "climate deniers!" The petition, full list and many interesting articles can be viewed at **petitionproject.org.**

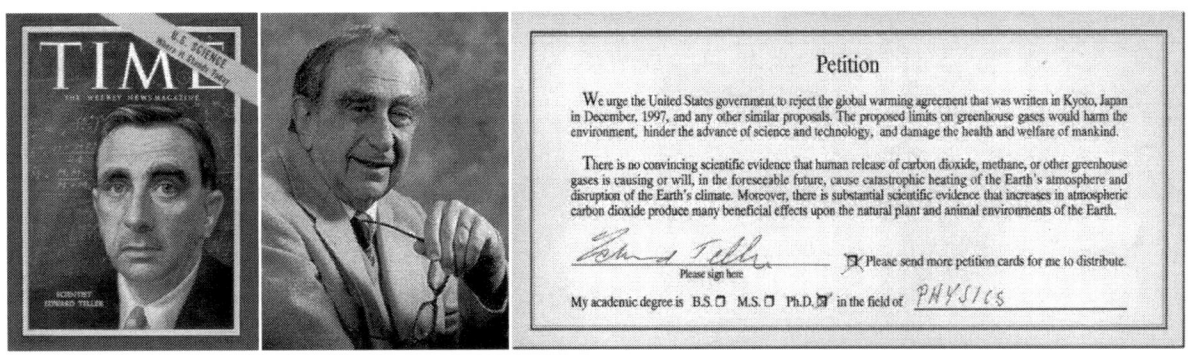

***Edward Teller** of Hydrogen Bomb fame once flirted with warmism, but then rejected it. Though deceased, he remains the most well-known of the 31,500 "denialist" signatories of the petition.*

2) The Debunking of the "97% Survey"

In 2013, an Australian warmist activist named **John Cook** and some associates published a paper claiming they had reviewed 12,000 abstracts of studies published in the peer-reviewed climate literature. Cook's team reported that 97 percent of the papers that expressed an opinion on man-made GW/CC *"endorsed the consensus position that humans are causing global warming."* As expected, Obongo and the Piranha Press took the "study" and ran with it. And thus was born the myth that "97% of scientists" are warmists.

The problem with Cook's rigged "study" is three-fold.

1) Cook, being a fanatical warmist, has a clear conflict of interest. How can any objective survey emerge from someone so heavily invested in "the cause?" Thus, we have no way of knowing if he cherry-picked the studies in such a way as to favor the warmist majority.

2) Because so much grant money and prestige is behind the propagation of GW/CC, a majority of studies that actually make it to publication and peer-review will, by necessity, be pro-warmist. The term used among academics is "publish or perish."

3) As was the case with other 'surveys' alleging an overwhelming "consensus" on GW/CC, the question surveyed had absolutely nothing to do with the debate between warmists and skeptics. The question Cook analyzed was simply whether humans have caused *some* degree of GW/CC. That question is meaningless because there are some skeptics do believe that humans *may* have caused some GW/CC, but that the contribution is so negligible that there it is nothing to worry about.

Investigative journalists at *Popular Technology* considered some of the papers which were classified within Cook's alleged 97%. They found that Cook and his crooked colleagues strikingly classified papers by outspoken "skeptics" such as **Willie Soon, Craig Idso, Nicola Scafetta, Nir Shaviv, Nils-Axel Morner** and **Alan Carlin** as supporting the 97-percent consensus!

Cook's crock of crap about "consensus" amounts to such a shoddy survey that even a few fellow warmists felt compelled to denounce its methodology. **Mike Hulme**, a Professor of Climate Change at University of East Anglia, had this to say about the 97%:

"The '97% consensus' article is poorly conceived, poorly designed and poorly executed. It obscures the complexities of the climate issue and it is a sign of the desperately poor level of public and policy debate in this country [UK] that the energy minister should cite it." **(21)**

And so, you see, dear reader, the most powerful "talking point" of the warmists, just like all the other baseless claims, is false. For those with a scientific/technical bent, a great place to gain better understanding of the full scope of scientific research to refute GW/CC is **ClimateDepot.com** -- an excellent resource that "they" don't want you to know about. Yes, indeed. The claim of overwhelming "consensus" is a monstrous lie. But that won't stop Fake Scientists, Fake Journalists and Fake Politicians from repeating it, over and over and over again.

*1. Obongo tweets out the 97% lie.   2. Scientist **John Coleman** -- who founded the Weather Channel -- refers to GW/CC as baloney. 3. In 2017, Professor **Judith Curry** -- former chair of the School of Earth and Atmospheric Sciences at the Georgia Institute of Technology -- announced the reason for her sudden resignation in a blog post:*

*"A deciding factor was that I no longer know what to say to students and post docs regarding how to navigate the CRAZINESS (caps hers) in the field of climate science. **Research and other professional activities are professionally rewarded only if they are channeled in certain directions approved by a politicized academic establishment** — funding, ease of getting your papers published, getting hired in prestigious positions, appointments to prestigious committees and boards, professional recognition, etc. --- How young scientists are to navigate all this is beyond me, and it often becomes **a battle of scientific integrity versus career suicide** (I have worked through these issues with a number of skeptical young scientists)."* **(22)** *(bold emphasis ours)*

"Career suicide" for opposing the sci-fi cult of warmism? A reasonable person would have to wonder, how many scientists who *claim* to believe in warmism are saying so for purposes of career advancement, or out of fear of committing "career suicide?" The cult of GW/CC may reward ambition and conformity and punish skepticism; but personal ambition and intimidation have no place in science.

INCONVENIENT TRUTH #8
PLANET MARS HAS BEEN WARMING UP LATELY

Oh those naughty little green Martians and their carbon-spewing SUV's!

Space.com *(May 26, 2016)*: **Red Planet Heats Up: Ice Age Ending on Mars**

"Mars is emerging from an ice age, a finding that could shed light on the past and future climates of both Mars and Earth, researchers said.

The orbit of Mars regularly undergoes changes that greatly affect how much sunlight reaches the planet's surface, which in turn can strongly alter the Red Planet's climate. Similar orbital variations called Milankovitch cycles are known to happen on Earth.

"All around the ice cap, there is evidence for a climate change from ice age to interglacial period," Smith (Isaac Smith, planetary scientist) told Space.com.

The researchers said their findings suggest that Mars recently emerged from an ice age, with ice beginning its retreat about 370,000 years ago.

These findings could help fine-tune Martian climate models, the researchers said. "We're still in the beginning phases of really understanding what happens in Martian weather patterns and longer cycles," Smith said. "Observations of current and recent processes keep giving us more information."

Moreover, these findings could help improve models of Earth's climate, he said. "Mars is relevant to Earth, because it has the same processes going on as Earth does, namely Milankovitch cycles," Smith said. "Mars serves as a simplified laboratory for testing climate models and scenarios, without oceans and biology, that we can then use to better understand Earth systems."

The scientists detailed their findings in the May 27 issue of the journal Science. **(23)**

The moral of the story? -- On other planets, as it is on Earth, there are far more significant factors, both known and *unknown*, which have a far greater impact on climate cycles than the relative *tiny* bit of CO_2 *(plant food)* than man's activities emit into the atmosphere.

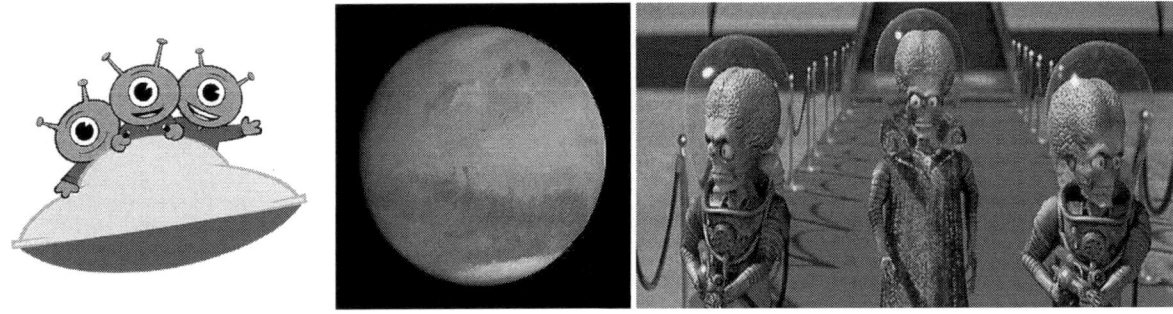

Martian spaceships are emitting too much CO_2.

INCONVENIENT TRUTH #9
MANMADE CO_2 ACCOUNTS FOR A TINY FRACTION OF "GREENHOUSE GASES"

Out of the entire atmospheric makeup, only about 1 is made up of "greenhouse gases." The other 99% is made up of nitrogen *(about 78%)* and oxygen *(about 21%)*.

Of *that* 1% percent, "planet-killing" carbon dioxide comprises less than 4% while water vapor encompasses 85-90%. Here is the estimated breakdown of the "greenhouse gas" quantities in parts-per-million *(of the total atmosphere)*.

Water Vapor: 12-15,000 ppm
Carbon Dioxide: 400 ppm *(man-made, 12 ppm)*

| **Methane:** | 1,800 ppm |
| **Nitrous Oxide:** | 300 ppm |

Total GG: *(low estimate)* 14,500 ppm

Manmade CO_2 therefore accounts for just 12 of the estimated 14,500 ppm of "greenhouse gas." So, if we kill every last human on Earth, we would only reduce GG to 14,488 ppm. Shouldn't we banning the water vapor, which outnumber manmade CO_2 molecules by a factor of 1000-1, instead?

The 5th grade math becomes even more ridiculous when we consider that water vapor and naturally caused CO_2 can fluctuate up or down by a few percentage points -- year-to-year, decade-to-decade, century-to-century. Therefore, a 1% rise in water vapor over a period of time would add up to an additional 120 ppm --- 10 times more ppm that the measly 12 we just eliminated by killing off 100% of humanity!

To better illustrate this in easy-to-grasp fact, imagine the Barclays Center Basketball Arena in Brooklyn, NY -- home to the Brooklyn Nets -- with its capacity of about 17,000 fans. Only about 15 of the 17,000 would represent manmade atmospheric CO_2 molecules, --- the rest being mainly water vapor, methane or natural CO_2.

It really is this simple. You see, basic 5th grade math, rooted in truth, trumps advanced level GW/CC physics rooted in bullshit.

In a packed arena full of greenhouse basketball "fans," only 15 of them were man-made molecules. Most were water vapor molecules.

The numbers get even more ridiculous. As previously stated, the global average concentration of CO_2 *(aka plant food)* in Earth's atmosphere is currently about 0.04%, or 400 *parts-per-million (ppm)*. In their own data charts, both the warmist US Department of Energy and the UN concede the fact that of that 400 ppm, just 3% is manmade -- with the other 97% coming from natural sources such as ocean outgassing, geothermal activity, volcanoes, naturally occurring forest fires, decomposition etc. **(24)**

Climate realist **Anthony Watts** quipped:

"If one wanted to make fun of the alleged consensus of "climate scientists", one could say that 97% of carbon dioxide molecules agree that global warming results from natural causes." **(25)**

Time for some more 5th grade math. To put this in easy-to-grasp perspective, imagine the University of Alabama's Bryant-Denny Stadium -- one of the largest football stadiums in America with a capacity of about 100,000 fans. Only 40 of the 100,000 would represent atmospheric CO_2 molecules, and only 1.2 people would represent the man-made CO_2 molecules. And the warmists expect us to believe that CO_2 is such a super-duper blocker/conductor of heat that only 1 "fan" added to the other 99,999 is going to melt the world's ice and deluge all of our major coastlines?

The very notion is preposterous on its face! Heck, even if 1 part per 100,000 of aerosolized aluminum *(a great conductor of heat)* was to be pumped into the atmosphere, it wouldn't make a damn bit of difference as far as temperatures were concerned. How could it?

"Yes, but the CO_2 accumulates," the ignorant warmist whines. Wrong! The plants eat it. More CO_2 means more and bigger green stuff. More green stuff means more CO_2 absorption *("carbon offset")*. If anything, we should try to deliberately *increase* the emission of this "pollutant" for a greener world and better crop production. But man's puny "carbon footprint" *(what a stupid term!)* couldn't even begin to keep up with nature's massive output of this natural plant food, even if he tried.

The 99,999 fans at the annual Auburn-Alabama football game were having a great time -- until he (T) showed up and melted Antarctica.

ANOTHER WAY TO LOOK AT THE NUMBERS:

The ratio of manmade CO_2 to the atmosphere is about the same as 8 gallons of water to a massive Olympic size swimming pool *(50 meters x 25 meters x 2 meters deep, 670,000 gallons)*. If a "scientist" was to claim that spraying 8 gallons of slightly heated water over the pool would cause the other 670,000 gallons to heat up, he would be laughed out of academia. So, why should we take warmists seriously for essentially suggesting the same thing with regard to CO_2? Heck! Even if the 8 gallons were *boiling,* there would be zero change in pool temperature.

vs ....

Silly warmist! How can 670,000 gallons possibly assimilate to the temperature of the added 8 gallons? Exactly the opposite would happen.

We have already established the fact that plant-food emissions, natural or manmade, are not causing the planet to heat up and the polar ice to melt. Nonetheless, just to cause some further discomfort and embarrassment to our warmist friends, we present this blast from the past from the "paper of record" that so many "intellectuals" worship with their morning lattes:

The New York Slimes *(October 31, 1982)*: **Termite Gas Exceeds Smokestack Pollution**

For several years scientists have been warning that carbon dioxide added to the atmosphere by increased burning of fuel is likely to alter world climates, like a greenhouse, by inhibiting the escape of heat into outer space.

*Now researchers report that **termites, digesting vegetable matter on a global basis, produce more than twice as much carbon dioxide as all the world's smokestacks.***

.... By digesting this debris, they are adding not only carbon dioxide but also methane to the atmosphere.

The high level of termite gas production is reported in the Nov. 5 issue of the journal Science. The authors measured termite gas production inside laboratory jars. In Guatemala forests, they enclosed a huge arboreal termite nest in a Teflon bag to confirm that the insects were prolific producers of methane.

*As pointed out Wednesday by one of the researchers, James P. Greenberg of the National Center for Atmospheric Research in Boulder, Colo., termites are far more abundant than most people realize. He estimated that there were **three quarters of a ton of termites for every person on earth.***

*Another author of the report, Patrick R. Zimmerman of the atmospheric center in Boulder, said that **plant respiration and decay added 10 to 15 times as much carbon dioxide to the air as termites.***

Other authors of the Science article were Dr. Paul J. Crutzen, director of the Max Planck Institute for Atmospheric Chemistry in Mainz, West Germany, and S.O. Wandiga of the University of Nairobi in Kenya. (emphasis added) **(26)**

The warmists dismiss the significance of these amazing termite statistics by claiming that although manmade emissions are relatively small, the natural balance is so delicate that our contribution "tips the scales." They not only offer zero observable, experimental evidence to support this claim, but they ignore the inconvenient truth of natural year-to-year CO_2 variations which can be greater than all of man's emissions. Evidently, the "delicate balance" ain't so delicate after all.

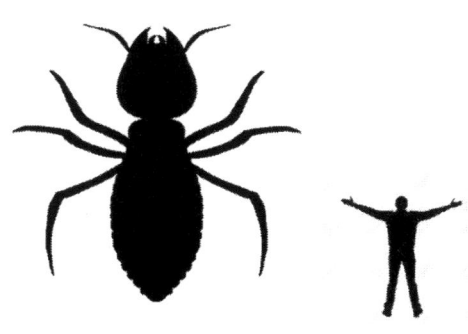

1. If we assume that the termite population has remained about the same, and factor in human population increase since 1982 (date of the article) there exist about 1/2 ton (1000 pounds) of termites per every human being-- all eating wood and emitting "greenhouse gas" day in and day out.
2. Perhaps the Terminix man can save the ice caps by spraying the planet's forests and jungles?

INCONVENIENT TRUTH #12
THERE ARE OTHER FACTORS THAT INFLUENCE CLIMATE FAR MORE THAN "GREENHOUSE GASES"

In addition to manmade CO_2 not even being a factor in retaining heat as a "greenhouse gas," the warmists generally downplay much more significant factors of what is referred to as "Earth's Energy Budget" as well as other variables that factor into the overall climate mix. These fluctuating and often unpredictable factors are what truly shape changes in climate -- not the measly 10-12 parts-per-million *(1 per 100,000!)* of manmade CO_2 which the crackpot warmists and their dupes speak of as if it were the very definition of climate.

Solar Radiation, Sunspots & Solar Winds:

Solar output is not constant. Sunspots are storms on the sun's surface marked by intense magnetic activity, solar flares and hot, gassy ejections. Cyclical Increases in sunspot activity have been observed to correlate to run-ups in global temperatures on Earth. Even the warmists concede that solar storms can impact climate. The only question is, how much?

Natural Axial Tilt Variation (Milankovith Cycles):

Though not part of "Earth's Energy Budget," **Milankovich cycles** describe the effects of changes in the Earth's movements on its climate. The term is named after Serbian geophysicist and astronomer **Milutin Milankovich**. In the 1920s, he theorized that variations in axial tilt resulted in cyclical variation in the solar radiation reaching the Earth. This influences climatic patterns on Earth. A trend of decreasing tilt leads to warmer winters and colder summers, as well as an overall cooling trend. A greater tilt makes the seasons more extreme.

Natural Changes in Wind Patterns

Wind patterns will vary only slightly from year-to-year, but are generally predictable within a given range. Over long periods of time, patterns can change substantially. Even the actual direction of "prevailing winds" can be altered significantly.

Science Daily *(January 24, 2007)*: Winds of Change: North America's Wind Patterns Have Shifted Significantly In the Past 30,000 Years

Summary:
> *Using 14,000- to 30,000-year-old wood samples from areas in the mid-latitudes of North America (40-50°N), researchers have learned that the prevailing winds in this region, which now blow from the west, once blew from the east.* **(27)**

Scientists can neither explain nor predict nature's adjustments, but obviously, when wind patterns vary, so does the weather. From a story about changing wind patterns affecting California:

UK Daily Mail *(September 24, 2014):* **Changing wind patterns, NOT global warming, are causing temperatures to rise on America's West Coast, says study**

"Increased temperatures on America's West Coast are not a result of human-caused climate change, but rather naturally occurring wind changes according to a new study.

The 1 degree Fahrenheit of warming that has been recorded on the coast of the northwest Pacific Ocean is due to weaker winds and changing ocean circulation, not a buildup of greenhouse gasses.

These weaker winds accounted for more than 80% of the warming trend along the Pacific Northwest coast between Washington and Northern California, and 60% of the warming in Southern California." **(28)**

Natural Changes in Ocean Currents

When those wind patterns change, the ocean currents can be affected -- and ocean current patterns affect climate in a big way. Consider how the Gulf Stream current, which emanates from the Gulf of Mexico, keeps the climate of England *(which is situated further north than Buffalo, NY, latitude-wise)* mild all year long.

Natural Changes in Cloud Formation

Because changes in solar radiation, wind patterns and ocean patterns can have significant atmospheric effects, cloud formation is also impacted. Unpredictable changes in cloud patterns and cloud "life-spans" affect rainfall and temperatures.

Variations in Geo-Thermal Heat

The interior of the Earth *(both the mantle and the crust)* generate what is known as "radiogenic heat." This type of heat can vary unpredictably from time to time and region to region. At the Earth's surface, if the incoming energy from the deep interior "furnaces" is greater than the outgoing energy flow, net heat is. These small variations, over time, along with pressure from the top, are what cause the gradual melting away of glaciers *from the bottom*. Place a chunk of ice on a slightly warm surface and observe how it melts faster from the bottom than at the top.

As a low ranking factor of "Earth's Energy Budget," radiogenic heat ranks far lower than whatever is happening with the Sun, Wind & Ocean patterns, and Axial Variation. Though it may not be a powerful net driver in shaping surface temperatures and climate; geothermal variations still account for a larger share of the climate mix than the puny effect of all greenhouse gasses, to say nothing of manmade CO_2.

Volcanic Emissions

Not part of the Energy Budget, the gases and dust particles spewed into the atmosphere during unpredictable volcanic eruptions *(or even just constant low-level activity)* can also influence climate. In the most extreme historical cases, the ash clouds were observed to spread across entire continents -- producing a dense haze that dimmed out the sunlight and caused extreme global cooling. **(29)** Conversely, periods of relatively low global volcanic activity, to the extent that they are not contributing to cooling, can correlate to "warming."

Cosmic Rays

Legendary French physicist **Joseph Fourier** *(1768-1830)*, believed that deep-space radiation was a significant factor which could influence climate variations. His theory on cosmic rays has since been vindicated by modern meteorology.

Principia Scientific: *(September 11, 2016)*: New Study: Solar & Cosmic Rays Impact Climate More than Expected

"Overview: It has long been widely accepted that the sun is absolutely critical to all weather and climate here on Earth and yet there are still some aspects of this connection that are not too well understood and even controversial.

For example, there has been the belief by many atmospheric scientists that cosmic rays which penetrate the Earth's atmosphere from outer space can play a significant role in the formation of clouds which, in turn, has a direct impact on climate. Solar activity has a direct impact on the ability of cosmic rays to actually reach the Earth's atmosphere. A just published study has confirmed the notion that cosmic rays can indeed be an important player in Earth's weather and climate and the role of the sun is critical." **(30)**

The X-Factors

It is certain that there are still other variables which impact climate that we haven't even discovered and may never discover. Philosophers understand and can accept the limitations of man's intellectual powers. Scientists, *particularly* the math-drunk "theoretical scientists," often cannot, or will not. The true believers among them -- as distinguished from the purposeful hoaxsters -- do not even pause to consider that they may only know a fraction of all that there is to be known. The ego-maniacs who truly think they've got climatology, or any other branch of science, all figured out and "settled" have already discredited themselves.

In the final analysis, manmade CO_2 represents only a tiny fraction of all atmospheric CO_2 ---which represents only a tiny fraction of the heat-retaining "greenhouse gasses" -- which represent just a tiny fraction of the Earth's Energy Budget -- which doesn't even include wind patterns, ocean current patterns, cloud patterns, axial tilt, volcanic particles and God only knows what else. And it is this fraction of a fraction of a 1% that the government-funded warmists insist will melt Antarctica and Greenland --- wiping out all of our coastal cities and towns. Give us a break!

The warmists prefer that we ignore this basic science and fixate on an insignificant bit of man-made CO_2 as the cause of a warming trend which isn't even happening at this time, natural or manmade. Interestingly enough, some warmists, when presented with actual data indicating no warming since 1998, will cite some of these factors as the reason why CO_2-induced warming is being "masked."

But when global temperatures do inevitably creep upwards *(as the result of natural cycles)* they'll be no more talk of solar storms, axial tilt, wind or current patterns. According to warmist dogma, any warming trend is the result of man-made CO_2. Any cooling trend must be the result of other factors. See how the trick works?

The fluctuations of solar radiation, wind power and direction, axial tilt etc. all affect the Earth's climate year-to-year, and century-to-century. Greenhouse gas is a bit player compared to those factors. And the small amount of manmade CO_2 is totally worthless as a warming factor.

INCONVENIENT TRUTH #13
RADICAL CHANGES IN CLIMATE HAPPENED LONG BEFORE THE INDUSTRIAL REVOLUTION

The use of the term "Climate Change" is dishonestly manipulative because it infers that climate would be static *(not changing)* were it not for man and his cars and smokestacks. Nothing could be further from the truth.

Again, apart from the fact that the planet is *not* even warming up at this point in time, the historical reality is that *pre-industrial* periods of sudden and extreme warming, and other *pre-industrial* periods of sudden and extreme cooling, are part of the natural cycles of climate. The warmists wish they could edit these periods out of history. Since they cannot erase the record, they generally ignore these events as best they can. Let's review a few of these "inconvenient" examples of pre-industrial or pre-automotive "climate change".

The 1920's Arctic Warming

We already covered this most recent case of radical warming and melting. It was less than 100 years ago, at a time when world population and world industry was just a fraction of what it is today. What caused this strange warming and melting of the Arctic Ocean! It wasn't man-made CO_2!

The Little Ice Age 1600s to 1800's

The most recent cool period, often called the "Little Ice Age," lasted about 200 years, finally ending in Western Europe around 1850. Amazing events that would be absolutely unimaginable today occurred routinely during those days. In 1658, a Swedish army marched across the frozen straights known as the "Great Belt" to Denmark to attack Copenhagen.

In 1794-1795 a French invasion army marched on the frozen rivers of the Netherlands, and a Dutch fleet got stuck in the ice in Den Helder harbor.

The sea ice surrounding Iceland extended for miles in every direction, closing harbors to shipping. Rivers in Great Britain were frequently frozen deeply enough to support ice skating and winter festivals. A Thames River "frost fair" was held from 1607 until 1814. **(31)**

The Medieval Warming Period

The warmest period of the last 2,000 years in the Northern Hemisphere occurred between 950 and 1100. Tree ring data shows that peak warmth occurred at different times for different regions, indicating that the Medieval Warm Period was not globally uniform.

Colonization of Greenland took place during this warm period as Vikings took advantage of ice-free seas to colonize areas in Greenland – some of which are still buried under ice today. What caused these areas of Southern Greenland to freeze up? Or, we might ask, what had caused the "greening" of Greenland before that? It wasn't man-made CO_2.

New York Times *(May 8, 2001)*: Story of Viking Colonies' Icy 'Pompeii' Unfolds From Ancient Greenland Farm

"At Nipaatsoq, blowing glacial sands covered the farm in the early 1400's, sealing it until 1990, when two hunters reported seeing ancient wood protruding from an eroded stream bank." **(32)**

Whenever any melting, due to natural variable factors, reveals some of these artifacts in Greenland or other places, the warmists are quick to shout, *"Aha! These artifacts are emerging because the ice is melting due to man-made GW/CC!"* But how is it that they never question why the areas were once warm and inhabited to begin with? And they also purposely ignore cases in which new ice-coverage / glaciers are forming.

The New Jersey Palisades

Now if you really want to go back in time to find some real "climate change," consider that the majestic smooth cliffs of the Hudson River Palisades. This line of steep cliffs *(300-500 feet high)* stands along the west side of the lower Hudson River in northeastern New Jersey and southeastern New York State. The formation itself was created by molten rock *(magma)* shooting upward, but the smooth polish is the effect of massive retreating glaciers which cut away the hillsides and loose rocks while sliding through. Imagine the prolonged cold spell which caused such skyscraper glaciers to form near today's midtown New York City. And imagine the relative warming that caused them to retreat and melt.

Needless to say, there were no cars and factories back then either.

1. A painting of the frozen-solid Thames River (London) Frost Fair. Natural warming put an end to the party after many years. ***2.*** *The New Jersey Palisades -- its hillsides were smoothed by sliding glaciers* ***3.*** *What exactly is "normal" climate anyway? The Medieval Warming Period flowed right into the Little Ice Age. --- NOTHING to do with CO$_2$.*

Judging by way that warmists speak of the "delicate balance" of Antarctic temperatures and the ice continent's associated "melting," one might think that average temperatures were hovering dangerously close to the water-to-ice "magic number" of 32 ° (F) / 0 ° Celsius. But just how cold is Antarctica?

With an ice-buried land mass about the size of the United States and Mexico combined, the mean annual temperature of the interior is *minus 70.6° / minus 57* °C. Satellite measurements have identified even lower ground temperatures of -135.8 °F / -93.2 °C at the East Antarctic Plateau. The coast is warmer and can go above freezing, but even the net amount of coastal ice *(the "shelf")* hasn't changed over the past 120 years.

In order to fulfill the doom & gloom false prophecies of inundated coastal cities and a submerged Statue of Liberty, the interior temps of Antarctica would need to *increase* by about **100 °F / 37 °C**. Heck! We in the temperate climate areas would all be burned to death anyway long before the "flood" described in Leonardo DiCaprio's propaganda film even arrived!

Only a stark raving lunatic "mad scientist" or a deceitful con-artist would argue, with a straight face, that Earth's climate will soon be like that of Planet Venus. Enter, from stage far-left, that "talking" stiff from *Weekend at Bernie's* said to be the "smartest man on Earth" -- Professor **Stephen Hawking** *(or is it his handler-ventriloquist talking?)*:

"We are close to the tipping point where global warming becomes irreversible. Trump's action (of pulling out of Paris Climate deal) could push the Earth over the brink, to become like Venus, with a temperature of 250 degrees, and raining sulphuric acid." **(33)**

Fear not, dear reader. Neither Antarctica nor the Statue of Liberty will be going away anytime soon.

The ass-clown "theoretical physicist" from "Weekend at Bernie's" (or his ventriloquist?) says that because of Trump, Earth will become like fiery Venus.

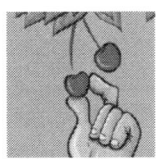 ***Cherry-Picked Data Alert!***

The warmists and the media are notorious for cherry-picking and breathlessly hyping-up cases of retreating *(melting)* glaciers as evidence of GW/CC. But what the warmists will not tell you is that there are glaciers which are currently *expanding*.

90

Here are a just a few of the typical contradictory headlines *(out of many!)* which illustrate how the scare tactic works:

- **On Shrinking "Melting" Glaciers:**

NPR *(National Public Radio, May 11):* **Disappearing Montana Glaciers a 'Bellwether' Of Melting To Come? (34)**

- **On Expanding Glaciers:** *(not nearly as publicized as the shrinking ones!)*

Live Science *(October 12, 2014)*: **Why Asia's Glaciers Are Mysteriously Expanding, Not Melting (35)**

Phys.org*: (February 15, 2017):* **Explaining New Zealand's unusual growing glaciers (36)**

*1. "Professor" Obongo lectures while on a hike to the Exit Glacier in Alaska -- a cherry-picked glacier that has actually been retreating since **BEFORE** the Industrial Revolution. **2.** Alaska's Mt. Hubbard's glacier is part of a long list of glaciers that are **growing** in locations on all continents. Obongo wasn't interested in any of those.*

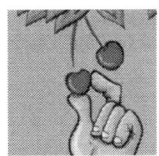 ***Cherry-Picked Data Alert!***

(Atolls are ring-shaped coral reefs created at the top of underwater volcanic accumulation. Coral are marine invertebrates. Their reefs include a coral rim that encircles a lagoon partially or completely. Over time, sand is deposited on the coral reefs, forming islands):

- **On Shrinking Atolls**

New York Time *(December 1, 2015)*: **The Marshall Islands Are Disappearing (37)**

Inverse.com *(May 9, 2016)*: **Six Pacific Islands Have Already Disappeared as Sea Levels Rise (38)**

- **On Expanding Atolls** *(not nearly as publicized as the shrinking ones!)*

The London Telegraph: *(June 2010)* **Low-Lying Pacific islands 'growing not shrinking' due to climate change (39)**

UK Daily Mail *(February 21, 2014)*: **The Island That 'Grew Back': Pacific Isle That Disappeared After Devastating Typhoon Reappears 100 Years After Its Destruction (40)**

New Scientist.com: *(June 2, 2015)*: **Small atoll Islands May Grow, Not Sink As Sea-Levels Rise (41)**

What the heck is going on here? Are glaciers retreating or are they expanding? Are atolls disappearing or are they growing? The answer is, *both* phenomena *are occurring* and *have always occurred* and *will continue to occur* in the future. Atoll erosion and restoration is related to the unpredictable cycles of tides and storms -- not to non-existent GW/CC! Similarly, Glacial shrinking or expansion is influenced by factors such as snowfall patterns, variations in Earth's internal heat, and just the natural cyclical variation in temperatures from one geographic area to another and from one period in history to another.

Of the many lies and omissions committed by warmist academics, politicians and journalists, this devious cherry-picking of shrinking glaciers and disappearing atolls represents perhaps the most blatant and easy-to-expose data manipulation tactic of the warmist Mafia. And it should not be misunderstood as merely "sloppy journalism" or "shoddy science." No sir. These sleazy sons-of-bitches know *exactly* what they are doing.

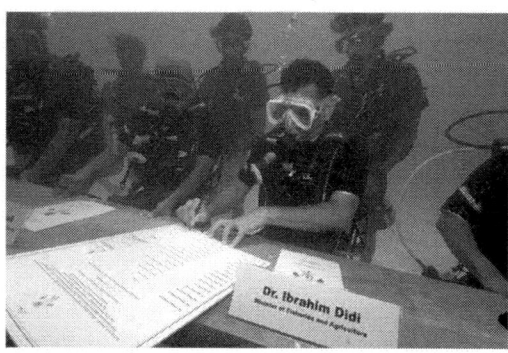 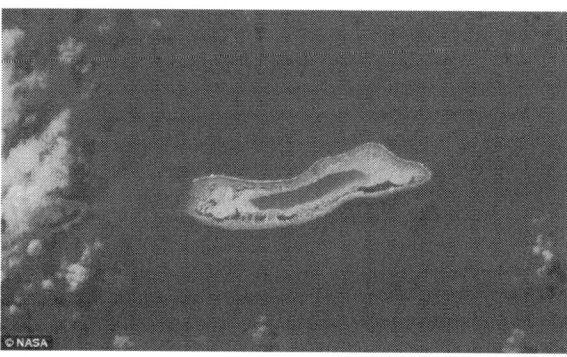

1. Crooked politicians of the low-lying Maldives islands hold an underwater cabinet meeting to demonstrate what a future of rising sea levels holds for their nation -- unless the United States gives them more money to build unnecessary seawalls and buy expensive new clothes and cars.
2. The Nadikdik Atoll actually grew back after eroding and disappearing due to a violent storm 100 years ago. Neither the disappearance nor the restoration had a damn thing to do with GW/CC.

INCONVENIENT TRUTH #17
WARMISTS SELECTIVELY USE DROUGHTS AND HEAT WAVES TO SCARE US ABOUT GW/CC

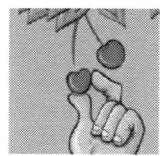 **Cherry-Picked Data Alert!**

One doesn't fully appreciate just how vast of a country the United States is until he drives across it or flies over it. And yet, as large as it is, the US represents only about 2% of the Earth's total surface territory, and about 8% of its land. And yet, the weather across just that small fraction of the Earth's surface can vary tremendously from New York to Atlanta; from Atlanta to Houston; from Houston to Chicago; from Chicago to Las Vegas; from Las Vegas to San Francisco; from San Francisco to Los Angeles and so on and so on. The point to understand here is that *at any given moment in time*, some major city or country on this huge planet of ours is experiencing a "100-year heat wave" or a "100-year drought" or a "100-year cold snap." It is a basic law of statistical distribution.

This statistical and geographic reality is what allows the warmists to cherry-pick and widely publicize inevitable extreme weather events year after year. The mathematically-challenged public will hear of a "100-year this" or a "100-year that" day in and day out, all year long, without ever realizing that the reporting of the extreme events, although *technically* true, only reflects a normal statistical distribution, not GW/CC.

The following headlines represent only a few typical examples of this shameless cherry-picking of data. Trust us, there are many, many more cases that can be found.

- **The 2000's Australian drought**

New York Times *(April 17, 2008)*: **A Drought in Australia, a Global Shortage of Rice**

The collapse of Australia's rice production may foretell some of the effects of global warming on agriculture **(42)**

- **The 2011-2016 The California Drought**

New York Times (August 20, 2015): **California Drought Is Made Worse by Global Warming, Scientists Say (43)**

- **The 2016 Indian heat wave**

New York Times (May 20, 2016): **Pray for Shade: Heat Wave Sets a Record in India .... "Climate change is obviously playing a role" (44)**

- **The Arizona heat wave of June 2017**

New York Times *(June 20, 2017)*: **Too Hot to Fly? Climate Change May Take Toll on Air Travel**

Excess heat in Phoenix grounded more than 40 flights in recent days -- Scientists say warming climate could mean more turbulent rides **(45)**

Each of the above-mentioned extreme weather events came and went -- just like countless historically recorded heat waves and droughts have come and gone since the days of antiquity. But the eventual passing of, and recovery from, these "100-year" events never get 1/1000th the publicity that the "disasters" themselves did when they first manifested. Be it the next heat wave, or the next drought, or the next hurricane, the next wave of Midwest tornadoes and so on -- the warmists stand ready to seize upon the next weather crisis somewhere else in this very big world and breathlessly declare, *"You see?! You see?!"*

Professor! Thou shalt not cherry-pick!

Do not "be worried" by cherry-picked data!

The USA is huge and full of many different climate regions. Yet it is small relative to the whole planet. So just imagine how many different possibilities can occur over the course of a year. There will always be some place experiencing extreme weather at any given time.

We now know that CO_2 is *not* an effective absorber / retainer of heat. We have also established that global temperatures are *not* rising at this time; and that the polar ice-caps are *not* melting away and that low-lying atoll islands and countries are *not* being inundated by "rising seas." And if those things were to happen someday in the future, the causes would be natural, not manmade. Therefore, the myth of rising sea-levels is a moot point that needn't even be discussed.

Nonetheless, just for the sale of thoroughness, let us kill this already killed lie even more. Excerpts from a UK Telegraph article on the findings of Swedish geologist and physicist **Nils-Axel Mörner**, a former chairman of the INQUA International Commission on Sea Level Change, give us the final nail in the coffin of GW/CC and its silly scare tactics of rising sea-levels threatening to swamp coastal areas, islands and even the Statue of Liberty.

UK Telegraph *(March 28, 2009)*: Rise of sea levels is 'the greatest lie ever told'

"The uncompromising verdict of Dr Mörner is that all this talk about the sea rising is nothing but a colossal scare story, writes Christopher Booker."

"If one thing more than any other is used to justify proposals that the world must spend tens of trillions of dollars on combating global warming, it is the belief that we face a disastrous rise in sea levels. The Antarctic and Greenland ice caps will melt, we are told, warming oceans will expand, and the result will be catastrophe.

Although the UN's Intergovernmental Panel on Climate Change (IPCC) only predicts a sea level rise of 59cm (17 inches) by 2100, Al Gore in his Oscar-winning film An Inconvenient Truth went much further, talking of 20 feet, and showing computer graphics of cities such as Shanghai and San Francisco half under water. We all know the graphic showing central London in similar plight. As for tiny island nations such as the Maldives and Tuvalu, as Prince Charles likes to tell us and the Archbishop of Canterbury was again parroting last week, they are due to vanish.

But if there is one scientist who knows more about sea levels than anyone else in the world it is the Swedish geologist and physicist Nils-Axel Mörner, formerly chairman of the INQUA International Commission on Sea Level Change. And the uncompromising verdict of Dr Mörner, who for 35 years has been using every known scientific method to study sea levels all over the globe, is that all this talk about the sea rising is nothing but a colossal scare story.

Despite fluctuations down as well as up, "the sea is not rising," he says. "It hasn't risen in 50 years." If there is any rise this century it will "not be more than 10cm (four inches), with an uncertainty of plus or minus 10cm". And quite apart from examining the hard evidence, he says, the elementary laws of physics (latent heat needed to melt ice) tell us that the apocalypse conjured up by Al Gore and Co could not possibly come about.

The reason why Dr Mörner, formerly a Stockholm professor, is so certain that these claims about sea level rise are 100% wrong is that they are all based on computer model predictions, whereas his findings are based on "going into the field to observe what is actually happening in the real world".

When running the International Commission on Sea Level Change, he launched a special project on the Maldives, whose leaders have for 20 years been calling for vast sums of international aid to stave off disaster. Six times he and his expert team visited the islands, to confirm that the sea has not risen for half a century. Before announcing his findings, he offered to show the inhabitants a film explaining why they had nothing to worry about. The government refused to let it be shown.

Similarly in Tuvalu, where local leaders have been calling for the inhabitants to be evacuated for 20 years, the sea has if anything dropped in recent decades. The only evidence the scaremongers can cite is based on the fact that extracting groundwater for pineapple growing has allowed seawater to seep in to replace it. Meanwhile, Venice has been sinking rather than the Adriatic rising, says Dr Mörner.

One of his most shocking discoveries was why the IPCC has been able to show sea levels rising by 2.3mm a year. Until 2003, even its own satellite-based evidence showed no upward trend. But suddenly the graph tilted upwards because the IPCC's favored experts had drawn on the finding of a single tide-gauge in Hong Kong harbor showing a 2.3mm rise. The entire global sea-level projection was then adjusted upwards by a "corrective factor" of 2.3mm, because, as the IPCC scientists admitted, they "needed to show a trend".

When I spoke to Dr Mörner last week, he expressed his continuing dismay at how the IPCC has fed the scare on this crucial issue. When asked to act as an "expert reviewer" on the IPCC's last two reports, he was "astonished to find that not one of their 22 contributing authors on sea levels was a sea level specialist: not one". Yet the results of all this "deliberate ignorance" and reliance on rigged computer models have become the most powerful single driver of the entire warmist hysteria." **(46)**

Other quotes from Dr. Morner:

"The sea is not rising. It has not risen in 50 years." **(47)**

"The late 20th Century sea-level rise lacks any sign of acceleration. Satellite altimetry indicates virtually no changes in the last decade." **(48)**

"You frighten a lot of scientists. If they say that climate is not changing, they lose their research grants. And some people cannot afford that; they become silent, or a few of us speak up because we think it is for the honesty of science that we have to do it." **(49)**

"Sea level rise does not exist in observational data, only in computer modeling." **(50)**

*Shhhhhh! You won't ever see a REAL scientist like **Dr. Mörner** featured in the New York Times or interviewed on "60 Minutes" or the "Charlie Rose Show." Why not?*

For marketing purposes -- and make no mistake, the entire GW/CC hoax is about marketing, not science -- the warmists have adopted to adorable big white bears of the Arctic as the "poster child" for their malevolent movement. As the fairy tale goes, the majestic white giants and their oh-so-cute little cubs are having a hard time making ends meet these days. With the ocean ice disappearing *(which it is not!)*, polar bears are being forced to swim further and further away from their historic ice-hole seal hunting grounds and turn to scavenging for survival.

Following is a typical example -- one of *thousands* -- of the manipulative heart-string-pulling nonsense about endangered polar bears being pumped out by the "elite" news media day after day.

New York Times / Editorial Board: *(December 20, 2016)*: **The Climate Refugees of the Arctic**

"The polar bear, the largest bear of them all and a fearsome predator, is the poster animal of climate change, and for good reason: While most threatened animals, such as the rhinoceros, are victims of localized threats like poaching or human encroachment, **the polar bear is threatened most gravely by global emissions of greenhouse gases** *(emphasis added). A polar bear was the star of Al Gore's celebrated 2006 film on climate change, "An Inconvenient Truth," and it has its own conservation organization, Polar Bears International, which has designated Feb. 27 as International Polar Bear Day."* **(51)**

The truth of the matter is that polar bear populations, according to several scientific studies, have *increased* over the past 12 years. Of course, the New York Slimes Editorial Board, led by the fanatical Marxist **Andrew Rosenthal**, is well aware of this fact -- which is why the writer felt compelled to vaguely allude to "uncertainties" before dismissing them. As the editorial summarizes the lies of a Slimes "science writer" named **Erica Goode**, observe the magician's clever little rhetorical move of casually mitigating a critical objection:

*"Using the bear as an icon to raise consciousness and funds ... does more than arouse support from conservationists. It also presents a ready target from climate change deniers who are only too willing to use **inevitable uncertainties** about the polar bear's actual numbers to challenge the facts of climate change."* **(52)**

Unfortunately, the significance of that split-second dismissive reference to *"inevitable uncertainties"* will be lost on most readers. Let's explore some of those *"inevitable uncertainties"* about the true state of the polar bear.--- shall we?

Daily Express (UK) *(February 28, 2015)*: **Polar bear populations are recovering despite the climate change warnings of environmentalists, a Canadian zoologist claimed yesterday.**

Dr. Susan Crockford said: "On almost every measure, things are looking good for polar bears."

In a report for the climate skeptic Global Warming Policy Foundation, she said: "Scientists are finding that polar bears are well distributed throughout their range and adapting well to changes in sea ice.

*"Health indicators are good and they are benefiting from abundant prey. **It really is time for the doom and gloom about polar bears to stop."** (emphasis added)*

Dr Crockford, of the University of Victoria in British Columbia added: "Polar bears are still a conservation success story. With a global population almost certainly greater than 25,000, we can say for sure that there are more polar bears now than 40 years ago.

"The global estimate is too high to qualify the polar bear as 'threatened' with extinction." **(53)**

Polar bears thriving despite the melting ice

Ben Webster

Polar bear populations are recovering well despite claims that declining Arctic sea ice is threatening their survival, according to a report by a group which disputes mainstream thinking on climate change.

There are at least 25,000 bears, more than double the number in the 1960s,

Polar Bear Science

← Recent studies show Sept ice of 3-5 mkm2 did not kill polar bears off as predicted

Climate change not forcing polar bears to hunt humans but lack of baby seals might →

Polar bear tragedy porn dressed up as science features in new BBC Earth video

Posted on September 17, 2016 | Comments Off

This new effort by the BBC would make the PR department of the Center for Biological Diversity proud, with it's prominent use of animal tragedy porn pretending to be science. In contrast, the actual science shows something quite different, though summer sea ice since 2007 has declined to levels not predicted until 2040-2070, there has been virtually no negative impact on polar bear health or survival, a result no one predicted back in 2005.

*Dr. **Susan Crockford** and many other researchers insist that polar bears are doing well. Why is it that their voices are never heard in the Piranha Press that professes to be "free?"*

From **PolarBearScience.com** *(December 23, 2015)*: **Survey Results: Svalbard polar bear numbers increased 42% over last 11 years**

*"Results of this fall's Barents Sea population survey have been released by the **Norwegian Polar Institute** and they are phenomenal: despite several years with poor ice conditions, there are more bears now (~975) than there were in 2004 (~685) around Svalbard (a 42 % increase) and the bears were in good condition.* **(54)**

Apart altogether from the facts that: 1) There is no observable long-term trend of retreating ice in the Arctic and 2) Polar Bears are doing just fine --- there is another annoying little *"inconvenient truth"* that, although seldom mentioned, is as undeniable as it is ironic. You see, boys and girls, because polar bears hunt for seals by breaking holes in the ice, thickening ice actually poses a serious threat to our furry friends to the North. Ice and snow which become too thick, due to extremely *cold* weather, becomes unbreakable. It's common sense. Again, **PolarBearScience.com** explains:

"Thick spring ice due to natural causes is currently the single biggest threat to polar bears. Not declining summer sea ice – thick spring ice. That could change in the future but right now, the evidence supports that statement.

*Marked **polar bear population declines have virtually always been associated with thick spring ice** that reduced local ringed seal prey."* **(55)** *(emphasis added)*

It is extreme cold, *not a bit of extra warmth*, that can make life miserable for Polar Bears. What irony!

That very same editorial also claimed that polar bears photographed scavenging whale bones and garbage dumps constituted clear evidence of a species in danger. Nice try at "Fake News," Rosenthal, but not exactly. Though certainly true that polar bears are talented seal hunters, they, like their Grizzly Bear and Black Bear counterparts, have *always* been known to scavenge for food as well. Only a dedicated propagandist or a blooming idiot would post manipulative images of scavenging bears as "proof" of GW/CC. If anything, excessive reliance on scavenging is a sign of too much *(too thick)* ice, not a shortage of it.

Such illogic is akin to claiming that someone has become a vegetarian because he was recently photographed snacking on a cucumber. But that's warmist logic for you -- the funniest, though one of the most dangerous, freak-shows on earth.

1. In cold weather, the bear says: "Bring some of that global warming over here. This ice is so darn thick that I can't bust through it to get my dinner!"
2 & 3. White or black, all bears will scavenge for food.

INCONVENIENT TRUTH #20
THE IMAGES USED TO FRIGHTEN THE PUBLIC ARE MISLEADING AND MANIPULATIVE

The purveyors of GW/CC propaganda understand all too well the old saying; *"a picture is worth a 1000 words."* And if that image is taken out-of-context or presented in a misleading manner, each of the "1000 words" constitutes a big lie that leaves a scary memory and a subconscious reaction which reason and logic will have a hard time dispelling. The three main propaganda images -- which we have all surely seen *countless* times -- that have been and still are used to shape "public opinion" are 1) Bellowing Factory Smokestacks 2) Stranded polar bears 3) Collapsing Antarctic glaciers.

Here is why you should not be frightened, at all, by these misleading made-for-TV images.

Smokestacks

Molecularly speaking, the smoke coming out of factory smokestacks is no different than the smoke emanating from millions of American, Canadian or European fireplaces at Christmas time; or the mouth-watering smoky smell from a summertime backyard barbecue; or the pleasing fragrance of an incense stick or decorative candle. As long as you aren't standing directly on top of the factory's chimney and directly inhaling, that smoke is essentially no different, no more "polluting," and no more dangerous than any other type of smoke. In short order, the smoke dissipates, the CO_2 content goes into the lower atmosphere, and plants "eat" it. Be not afraid of the illusion!

Don't allow the warmists to blow scary smoke up your butt. It's no different than your own home chimney or backyard BBQ.

Stranded or Swimming Polar Bears

The next time you see an "heart-breaking" image of a polar bear "stranded" on a piece of floating ice in the "disappearing" Arctic, or swimming far from shore, understand that he is doing exactly what he was made to do --- nothing more. Because their buoyant blubber helps them to float very easily, polar bears regularly swim up to 40 miles, and have been recorded swimming as far as 400 miles!

National Geographic *(July 22, 2011)*: Longest Polar Bear Swim Recorded— 426 Miles Straight (56)

And GPS tracking devices have revealed that polar bears can swim, non-stop, for 10 days straight! **(57)** At just 3 months of age, even the little cubs can swim! **(58)** Weeping for a swimming polar bear is like feeling sorry for a floating duck. Be not afraid of this illusion either!

Sad pictures from Al Gore's "Inconvenient Truth" leads fools to believe that the bears (who can swim for hundreds of miles at a time!) are helplessly stranded.

Collapsing Polar Glaciers & New Icebergs

 The heavy and unstable floating ice cliffs *(known as ice shelves)* extending out from the land perimeter of Antarctica or Greenland will often chip off and break, forming icebergs. Thousands of icebergs break away from the protruding shelves each years --- some small, some as big as American cities! This process is referred to as "ice calving." Ironically, calving generally results from *expanding* ice accumulation and accumulated pressure, not "thawing!" Leave it to the slippery, sleazy, slimy warmists of academia and the media to misrepresent what is usually a sign of *cooling* as an example of warming!

Meanwhile, along another given portion of the edge, other glacial ice cliffs are expanding. Far from being a net loss of ice along the perimeter, the Antarctic ice coverage, as ironically confirmed by NASA's satellite data, has been slightly expanding. Old ice breaks. New ice forms --- all part of the natural cycle. And besides, collapsing ice that is already on top of water wouldn't increase sea-levels anyway. Again, try melting some ice cubes in a glass of water and watch the liquid level stay the same or actually *decrease* slightly as the floating ice melts. It would only be land-based melting ice flowing into the sea that would add volume to the sea. As for the land interior of the ice-buried mass continent of Antarctica, well, with average temperatures of - **70 F**, that ice ain't goin' anywhere!

Lose no sleep over scary-looking videos of collapsing blocks of floating ice, or stories of massive icebergs chipping off of the shelves of Antarctica or Greenland. That's an out-of-context illusion too.

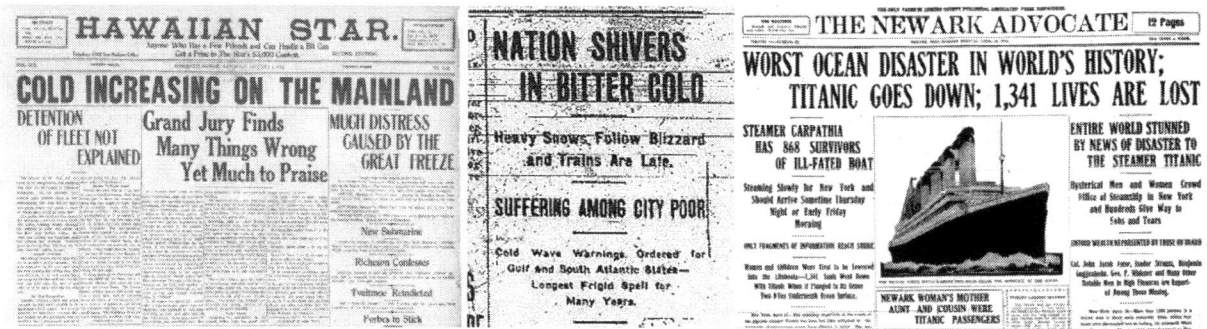

The above headlines are all from 1912.

The cold wave of 1911-1912 remains the coldest winter ever recorded in the US and Canada. **(59)** *It is no coincidence that the Titanic struck an iceberg in April of 1912 because it is extreme cold and expanding ice --* **not** *"thawing" – which accelerates the natural process of "ice calving."*

INCONVENIENT TRUTH #21
AL GORE'S FAMOUS FILM -- "AN INCONVENIENT TRUTH" -- WAS FOUND BY A BRITISH COURT TO HAVE BEEN RIDDLED WITH NINE MAJOR ERRORS

When the British Department of Education tried to force former US Vice President Al Gore's hyped-up propaganda film, *An Inconvenient Truth*, on students, a British parent sued to stop the attempted brainwashing. The court found, and the makers of the film agreed, that there were at least nine substantive errors in the film, most notably the notorious and thoroughly debunked **(60)** "Hockey Stick" temperature graph. Though the court did not ban the film, it was agreed that teachers would be required to point out the errors to their captive students.

In 2009, film-makers **Ann McElhinney** and **Phelim McAleer** released *Not Evil Just Wrong* -- a documentary which highlights not only the gross errors *of An Inconvenient Truth*, but also other facts that the warmists conveniently like to omit from their false narratives--- such as the Medieval Warming Period of 1000 years ago.

The debunking film can be purchased at **NotEvilJustWrong.com** or found on YouTube. It's a great documentary, but we cannot agree with the title. The producers, obviously afraid of sounding *too* "controversial," insist that the warmists are "not evil" --- but "just wrong." A better, more accurate title would have been *"Not Just Wrong, But Also Evil."* As we shall see in Section 5, notwithstanding the dupes of the GW/CC conspiracy, the ultimate controllers have evil motives for pushing this crap as hard as they have been.

1 & 2. "Not Evil, Just Wrong" tells the amazing story of how a British court trashed "An Inconvenient Truth" -- the "crockumentary" film which won Al Gore a Nobel Prize, an Academy Award, and a media puff-up as a "thinking man." *(Image 3)*

INCONVENIENT TRUTH #22
THE "CLIMATEGATE" SCANDAL OF 2009

If anything should have killed the GW/CC hoax once and for all time, it should have been the infamous scandal dubbed **"Climategate"** (*a name play of the Watergate Hotel Scandal of 1972-73 which took down President Richard Nixon*).

Climategate began in November 2009 with the hacking of a server at the Climatic Research Unit (CRU) at the University of East Anglia (UK) by an external attacker of unknown origin. Thousands of emails and computer files were copied and sent to various internet locations several weeks before the Copenhagen Summit on climate change. Copies were sent to the all-mighty all-Globalist BBC *(British Broadcasting Corporation)*, which dutifully ignored them.

The story was finally made public by a fairly well-known British novelist and journalist named **James Delingpole**. The E-mails made it clear, *beyond any doubt*, that some of the top warmist scientists in the world were openly conspiring to manipulate climate data, to bar critics from accessing their records, and to influence scientific publications to censor "climate deniers."

Partly because Delingpole was a columnist of some significance, and mainly because the E-mails themselves -- in which top warmists openly spoke of using *"Mike's (math) trick to hide the decline" (in recent temperatures);* concealing data from a Freedom-Of-Information-Act request; censoring critics; complaining about how the reality of the Medieval Warming Period undermined their claims; and even welcoming the recent death of a well-known "denier" -- were all so outrageous that some of the "mainstream media" and even the UN was forced to confront the "Climategate" scandal -- albeit in a 'limited hangout" sort of way.

The ensuing official reaction to the scandal entailed a gentle wrist-slapping of those involved along with a public "laying low" of the hoaxsters for a year or two. The public and the sheep-like "scientific community" were reassured that although some of the warmist E-mails were "unfortunate," it was all "taken out of context" and really not that big of a deal. The fairy tale of GW/CC is still valid. Everybody move along now -- nothing to see.

The best and easiest-to-understand synopsis of the scandal is Meteorologist Brian Sussman's *book Climategate: A Veteran Meteorologist Exposes the Global Warming Scam.* **Sussman's** riveting account will not only explain the significance of the hacked E-mails, but also school you on the pseudo-science of GW/CC.

1. *There were some British newspapers -- which evidently still maintain some independence – that covered the Climategate scandal. In America, coverage was minimal.* **2&3**. *American Meteorologist Brian Sussman – His book "Climategate" is a devastating expose of the scandal.*

INCONVENIENT TRUTH # 23
49 FORMER NASA SCIENTISTS AND ASTRONAUTS SIGN OPEN LETTER CONDEMNING NASA FOR ITS INACCURATE WARMISM

The **NASA** name-brand commands respect and awe for its scientific achievements, mainly in the fields of rocket science and space exploration. Because of that, the agency, *(full name is **National Aeronautics and Space Administration**)* which is part of the Executive Branch Federal government, has been harnessed for political purposes.

Prior to the warmist scam, the single most disgraceful example of the presidential politicization of NASA occurred in 1990. President **George Bush the 41st** and his inner circle were looking for a phony pretext to attack Iraq and weaken its ruler, **Saddam Hussein**. After essentially baiting Iraq to invade the tiny state of Kuwait *(formerly a part of Iraq)* the Bush administration falsely claimed that Iraq was preparing to invade Saudi Arabia next. To support this monstrous lie, it was announced by Bush's henchmen *(and breathlessly repeated by the complicit media)* that NASA had satellite imagery showing a massive formation of Iraqi troops on the Iraq-Saudi border. To make a long story short: after the brief war, neither the original source for the claim nor the actual NASA satellite photos could be produced. NASA would neither confirm nor deny the original claim -- thus making the agency criminally complicit in deceiving the American people into an unnecessary war. **(61)**

And so, if historical precedent is any indication, we know that if a warmist US President like Clinton or Obongo wants to push the GW/CC hoax, the politically-appointed head of NASA *(and its "Earth Science" division)* will see to it that it gets done. The "you-know-what", as they say in the corporate world, will roll downhill from there, and career-minded scientists and bureaucrats on the lower levels know to either get on board the GW/CC Express, or at least keep their doubts to themselves.

In 2012, an open-letter was released which should have been covered by every major newspaper and TV News program in America. But it was ignored for reasons that we'll reveal in Section 5. An impressive group of 49 scientists and astronauts *(all ex-NASA)* -- featuring such distinguished names as **Michael F. Collins, Walter Cunningham**, five Apollo astronauts, and two former directors of NASA's Johnson Space Center in Houston -- sent and announced the open-letter to NASA President and Obongo appointee **Charles "Charlie" Bolden**. The strongly worded letter included a request for NASA to refrain from mentioning CO_2 as a cause of global warming in future press releases and websites.

Here it is:

The Honorable Charles Bolden, Jr.
NASA Administrator
NASA Headquarters
Washington, D.C. 20546-0001

Dear Charlie,

We, the undersigned, respectfully request that NASA and the Goddard Institute for Space Studies (GISS) refrain from including unproven remarks in public releases and websites. We believe the claims by NASA and GISS, that man-made carbon dioxide is having a catastrophic impact on global climate change are not substantiated, especially when considering thousands of years of empirical data. With hundreds of well-known climate scientists and tens of thousands of other scientists publicly declaring their disbelief in the catastrophic forecasts, coming particularly from the GISS leadership, it is clear that the science is NOT settled.

The unbridled advocacy of CO2 being the major cause of climate change is unbecoming of NASA's history of making an objective assessment of all available scientific data prior to making decisions or public statements.

As former NASA employees, we feel that NASA's advocacy of an extreme position, prior to a thorough study of the possible overwhelming impact of natural climate drivers is inappropriate. We request that NASA refrain from including unproven and unsupported remarks in its future releases and websites on this subject. At risk is damage to the exemplary reputation of NASA, NASA's current or former scientists and employees, and even the reputation of science itself.

For additional information regarding the science behind our concern, we recommend that you contact Harrison Schmitt or Walter Cunningham, or others they can recommend to you. Thank you for considering this request.

Sincerely, (Attached signatures (49 of them)) **(62)**

Oh well, so much for that "consensus" of "settled science!" But instead of giving the ex-NASA scientists any attention, the media continued to trot out High School teacher clowns like **Bill Nye** the "Science Guy" or the buffoonish **Neil DeGrasse Tyson**. The open-letter makes it very clear: NASA has become political, which makes it an unfit "expert witness" in the trial of GW/CC.

*1. In 1990, Bush 41 and his Secretary of Offense **Dick Cheney** used NASA's name to sell a lie. 2. In 2009, Obongo, the first Black President, appointed Bolden, to be the first Black Chief Administrator of NASA. Bolden obeyed his master and kept the GW/CC propaganda going strong.*
3. NASA should return to its proper function of rocket and space science.

INCONVENIENT TRUTH #24
NOAA HAS ALSO BEEN POLITICIZED

Along with their co-warmists at **NASA**, the Federal agency known as **NOAA** *(National Oceanic and Atmospheric Administration)* is considered the other main "go to" source for warmist data. And just like NASA, NOAA has also been rocked by internal dissent over rigged data and unscientific warmism. To give you an idea of just how important NOAA is to the warmists, consider that ex-President Obongo announced that he was going to nominate outspoken warmist **Jane Lubchenko** to be the new chief administrator for the agency -- a full month *before* he was even inaugurated as the 44th President. Lubchenko stepped down in 2013 to return to academia. She was succeeded at NOAA by another fanatical female warmist named **Kathryn Sullivan**, who stepped down after the election of Donald Trump.

*1. **Jane Lubchenko** of NOAA and Oregon State studies the melting (or is that expanding?) ice **2**. Kathryn Sullivan of NOAA gets chummy with Affirmative Action made-for-TV science clown **Neil DeGrasse Tyson**. **3**. NOAA pushes fake science*

Given what we know about the fanaticism, intolerance and dishonesty of the warmists, it would not be at all surprising, but rather *expected*, that heavily politicized NOAA would be manipulating data in order to "prove" GW/CC. In February, 2017, we received confirmation of this well-grounded suspicion from **Dr. John Bates**, a high-level whistle-blowing scientist within NOAA.

From the UK Daily Mail *(February 4, 2017)*: Exposed: How world leaders were duped into investing billions over manipulated global warming data

"The Mail on Sunday today reveals astonishing evidence that the organization that is the world's leading source of climate data rushed to publish a landmark paper that exaggerated global warming and was timed to influence the historic Paris Agreement on climate change.

111

*A high-level whistleblower has told this newspaper that America's National Oceanic and Atmospheric Administration **(NOAA) breached its own rules on scientific integrity when it published the sensational but flawed report, aimed at making the maximum possible impact on world leaders** at the UN climate conference in Paris in 2015.*

*The report claimed that the 'pause' or 'slowdown' in global warming in the period since 1998 – revealed by UN scientists in 2013 – never existed, and that world temperatures had been rising faster than scientists expected. **Launched by NOAA with a public relations fanfare, it was splashed across the world's media, and cited repeatedly by politicians and policy makers.***

But the whistleblower, Dr John Bates, a top NOAA scientist with an impeccable reputation, has shown The Mail on Sunday irrefutable evidence that the paper was based on misleading, 'unverified' data. It was never subjected to NOAA's rigorous internal evaluation process – which Dr Bates devised.

*His vehement objections to the publication of the faulty data were overridden by his NOAA superiors in what he describes as a **'blatant attempt to intensify the impact'** of what became known as the "Pausebuster" paper.*

His disclosures are likely to stiffen President Trump's determination to enact his pledges to reverse his predecessor's 'green' policies, and to withdraw from the Paris deal – so triggering an intense political row.

*In an exclusive interview, Dr Bates accused the lead author of the paper, Thomas Karl, who was until last year director of the NOAA section that produces climate data of **'insisting on decisions and scientific choices that maximized warming and minimised documentation… in an effort to discredit the notion of a global warming pause**, rushed so that he could time publication to influence national and international deliberations on climate policy'.*

Dr Bates was one of two Principal Scientists at NCEI, based in Asheville, North Carolina." **(63)**

The revelation that an inside whistle-blower claimed that rigged numbers were used to influence the delegates at the historic Paris Climate Conference is a blockbuster allegation that should have been plastered across the front page of every major newspaper in America *(and the world!)* and been the lead story on every one of those superficial 30-minute infomercials known as "the Nightly

News." There should have been Senate hearings, Congressional hearings, subpoenas and, if found to be true, lots of arrests and convictions!

Instead, with few exceptions, Dr. Bates was treated as though he never existed. That alone is evidence of a massive worldwide conspiracy which, we state again, will be explored and exposed in Section 5.

*1 & 2. According to **Dr. John Bates** of NOAA, rigged numbers from NOAA were used to wrongly discredit "the pause" and to prop up the crooked Paris Climate Agreement of 2016 (signed in 2017) 3. A Paris climate-activist adds outside pressure to any wavering delegates. Though some are true-believing fools, most "climate activists" are also phonies. Many of them are paid to demonstrate and others are admitted communists.*

INCONVENIENT TRUTH #25
THE "SCIENCE" OF GW/CC IS STEEPED IN MONEY AND POLITICS

The old biblical adage about money has been repeated so many times that it has become a cliché. You know: ***"The love of money is the root of all evil."***

Another worn out "oldie but goodie" warning pertains to power -- which is really just another word for politics. You've heard it: ***"Power corrupts and absolute power corrupts absolutely."***

Indeed, money and power are the two most corrupting influences known to man. Even many a good man has succumbed to their corrupting temptations, particularly if his livelihood depends upon grant money. In much the same way that a "starving artist" may sell her body to a Hollywood producer *(the proverbial "casting*

113

couch"), a starving or ambitious climatologist, theoretical physicist or oceanographer will sell his creative "research" and math skills to a higher power.

But true science represents a quest for truth and nothing but the truth. Should even the slightest bit of moral corruption enter into the process, science can quickly lose its vital connection to objectivity and turn into science-fiction. When we examine this weird obsession to promote GW/CC, we find that it is absolutely *steeped* in two things: **money and politics!**

Virtually all of the generous grants **(money)** awarded to researchers and universities for the "study" of GW/CC either comes from a government entity or a Left Wing foundation **(politics/power).** One would be hard-pressed to find *any* GW/CC "research" or "polar expedition" that *isn't* funded by one of these two sources. That alone should be enough to put any rational thinking person into red-alert skepticism mode.

The US Federal Government, through NASA, NOAA, the EPA and the Energy Department has doled out billions in grant money earmarked *only* for warmists. Other major government funders of various GW/CC studies and initiatives include the massive US state of California, the national governments of just about every liberal western country, the European Union, NATO, the IMF, the World Bank and, of course, the United Nations.

Left-Wing Globalist "think tanks" and foundations also pump billions of dollars into the GW/CC gravy trains. Here is just a *partial* list of some of the Globalist "progressive" heavies showering the warmists with never-ending boatloads easy cash:

Annenberg Foundation, Carnegie Foundation, Crown Family Philanthropies, Doris Duke Charitable Foundation, Ford Foundation, Bill & Melinda Gates Foundation, Hewlett Foundation, MacArthur Foundation, McKnight Foundation, Mitchell Foundation, Packard Foundation, Rockefeller Brothers Fund, Rockefeller Family Fund, Rockefeller Foundation, the various fronts of George Soros and the Ted Turner Foundation.

2001 / Carnegie Medal winners for "philanthropy"

*Standing: left to right: Billionaire Boys **Ted Turner, Bill Gates Sr., George Soros, David Rockefeller**. Seated: left to right, Billionaire Girls **Irene Diamond, Leonore Annenberg, Brook Astor.** These oh-so-generous elites have poured many, many millions of dollars into funding whore-scientists to promote the GW/CC hoax. They are also huge supporters of the United Nations. What is the real reason for such "generosity?"*

Apart from the fortune to be had from pro-Globalist governments and pro-Globalist foundations, there is more easy money to be made from peddling GW/CC alarmism books or "crockumentaries." Any well-connected "scientist" who writes a book on the "crisis" will dutifully be puffed-up by the **New York Slimes,** the **Washington Compost,** and PBS talk-show charlatan **Charlie Rose** into best-seller status. But dare to question the holy dogma, and be prepared to self-publish and starve.

And then, on top of that, the New York Slimes, quoting a former US State Department "climate adviser", reveals:

New York Times *(July 8, 2017)* World Leaders Move Forward on Climate Change, Without U.S.

*"the clean energy marketplace created by the Paris Agreement is estimated to be worth over **20 trillion dollars**." (64) (emphasis added)*

With **20 trillion** dollars of state-coerced "green energy" profits to be raked in by the well-connected, there are plenty of high-powered people -- such as billionaire solar tycoon, **Elon Musk** -- who will have much to gain from "saving the planet" with impractical "green energy" sources that are not as nearly as efficient as the **harmless** burning of coal, oil and natural gas.

Here's another proverb: *"He who writes the checks, calls the shots."* Be patient, boys and girls. We shall address the higher motives of these high-fallutin' "check-writers" in Section 5. For now, all you need to understand is, as previously stated: both the "science" and the "green energy solutions" to GW/CC are completely -- and we do mean *completely* -- steeped in the most corrupting influences known to man -- money and politics.

The **REAL** mathematics of GW/CC: BM + BP = FS

(Big Money + Big Politics = Fake Science)

SECTION 3

THE FAKE LOGIC OF GLOBAL WARMING / CLIMATE CHANGE

"Those who are able to see beyond the shadows and lies of their culture will never be understood, let alone believed, by the masses." (1)

- *Plato*

THE "SCIENTIFIC" AND RHETORICAL ARGUMENTS FOR GW/CC ARE BASED ON THE CLASSIC LOGICAL FALLACIES DEFINED BY GRAECO-ROMAN PHILOSOPHY

In studying the arguments in favor of man-made GW/CC, the student of classical philosophy cannot help but be thunderstruck at both the amount and the severity of the classic logical fallacies spewed forth by the warmists. A logical fallacy is a flaw in reasoning. They are *illusions* of thought, and they can be very tricky.

These fallacies can result from deliberate dishonesty, psychological insecurities *(over being wrong)* or just a partial or total lack of true education. Warmists in government, academia and media generally fit into one of those categories.

Put your science mindset away for now. Philosophy class is now in session.

The Appeal to Authority Fallacy

Example: Professor Mann is an expert on climatology and he says that manmade CO_2 will lead to catastrophic Global Warming

Why is this fallacious? The argument is not based on any facts, but solely on an appeal to the alleged "authority" of Professor Mann. It proves nothing.

The Appeal to Motive Fallacy

Example: Senator Paul opposes the regulation of CO_2 because he represents the coal-mining state of Kentucky.

Why is this fallacious? The argument is not based on any facts, but solely on ascribing a potential motive for Senator Paul's "denial." It could very well be that Senator Paul understands certain facts that lead him to conclude that CO_2 is not a problem, after all.

The Ad Hominem Attack Fallacy

Example: The use of the idiotic term "Climate Denier" or the dismissal of non-believers as "conspiracy theorists," "uneducated" or "crazy."

Why is this fallacious? The insults are not based on any facts. It might be perfectly acceptable to call a "denier" stupid or "crazy" if the case for GW/CC were clearly proven – but it simply has *not* been proven.

Instead of presenting actual scientific evidence to prove their fake theory, the warmists discredit real scientists by calling them "climate change deniers." This is propaganda, not science.

Argument from Fallacy

Example: When climate realist Senator **Jim Inhofe** (R-OK), in a misguided attempt to mock the warmists, brought a snowball to the Senate from an unusual springtime snowstorm that had struck Washington, DC, the warmists were quick to correctly frame Inhofe's stunt as an example of someone mistaking weather and climate. Inhofe *(who, ironically, is very well informed on the subject of fake GW/CC)* was mercilessly mocked for days by the warmist press. By extension, all "climate deniers" were also portrayed as simple-minded fools who ignorantly conflate day-to-day weather with long term climate trends.

Why is this fallacious? The warmists tactical fixation on a single logical fallacy committed by Senator Inhofe, does not in any way negate the mountain of other evidence which destroys their fake science of man-made GW/CC.

The Fallacy of Verbosity and Complexity

Example: The wordy technical explanations offered for each and every minor detail of GW/CC combined with the theoretical physics and computer models.

Why is this fallacious? This tactic serves to confuse and intimidate those without a technical background, while dazzling those with a passion for math and physics. The fallacy lies in the fact that there is no provable connection between the complex calculations and the physical reality. A complex veneer of science, as brilliantly concocted as it may be, should not be mistaken for actual science.

The Appeal to Force Fallacy

Example: Many non-believers now face "career suicide," or at least a professional setback, if they dare to express their disbelief. This is particularly true among younger, up and coming scientists. Another variation of this tactic is the *mandatory* GW/CC brainwashing of young students from Middle School up through High School

Why is this fallacious? It uses force rather than facts and persuasion to "educate" non-believers.

The Ad Populum Fallacy

Example: The claim that there is a "consensus" among scientists and/or the general public that man-made GW/CC is happening.

Why is this fallacious? Apart from the fact that this is a damn lie, the argument is not based on any facts, but solely on assumption that the majority is always right.

Lying

Example: The manipulation of data uncovered by the "Climategate Scandal" or the convenient "adjustment" of satellite RSS data.

Why is this fallacious? This one is self-explanatory. Suffice it to say that once a person is caught lying, a reasonable man has every right to disbelieve and disregard the liar's other claims too.

Ninety-seven percent of scientists agree: #climate change is real, man-made and dangerous. Read more: OFA.BO/gJsdFp

← Reply 🔁 Retweet ★ Favorite ••• More

*Can you spot the **three** fallacies contained in Obongo's single "tweet?"*

***1.** Appeal to Authority of "scientists" **2.** Ad Populum Appeal to the 97% majority   **3.** The 97% claim is a blatant **lie** concocted by a green activist and a rigged "survey."*

The "Fact Check" Trick

Example: A warmist journalist or a left-wing group *(such as Snopes, FactCheck.org, Skeptical Science etc)*, pretending to be "objective," will "analyze" a true claim which contradicts a warmist lie, and then falsely declare: *"We "fact-checked" this claim and it turns out that it is wrong."*

Why is this fallacious? Fake "fact-checking" is just a sophisticated form of lying -- packaged in such a way that it appears objective, authoritative and definitive. Warmists use this little trick very frequently.

Syllogistic Fallacy / Existential Fallacy / Prior Assumption

Example: Headline: New York Times, *(June 20, 2017)*: Too Hot to Fly? Climate Change May Take Toll on Air Travel (2)

Notice how the headline just automatically assumes, *as a given*, that GW/CC is occurring, and then links it air travel cancellations *(due to a Phoenix heat wave)*.

Another common variation of this fallacy among warmists is employed when they are challenged over the lack of warming.

Example: *"The reason why we are not seeing warming now is unclear, but it may be due to (fill in the excuse)."* The obvious fallacy is that the option of man-made CO_2 being a false theory to begin with is never considered as a possible reason for the lack of any substantive warming trend. You see, GW/CC is *always* fallaciously accepted as a given foundation for all subsequent arguments.

Why is this fallacious? The prior assumption of man-made GW/CC has not been established by facts and observation. By constantly slipping the lie in there as a prior assumption, the reader is left with the false impression that it has been previously proven to be true.

The Appeal to Emotion Fallacy

Example: The use of cute polar bears and polar bear cubs as the face of GW/CC

Why is this fallacious? Who doesn't love bear cubs? The idea that cuddly little cubs are dying out due to manmade CO_2 causes soft-hearted people to lose their sense of judgment. No facts here --- just a sneaky tugging of the heart strings.

The Anecdotal Fallacy (Cherry-Picking)

Example: The claim that a particular country just experienced its hottest summer in 100 years.

Why is this fallacious? On this huge planet of ours, there will always be a region that just experienced unusual warmth, and there will simultaneously be regions that are experiencing unusual cold. Cherry-picked events prove nothing.

The Burden of Proof Fallacy

Example: The warmist will challenge the "deniers" to disapprove a particular claim about their theories regarding ice core samples and tree rings.

Why is this fallacious? The burden of proof is on the person making a claim, not on someone else to prove it false. Who among us, if falsely accused by a clever theorist of committing an unsolved murder many years ago, would be able to *disprove* the theory? Nonetheless, most of the warmists claims have been disproven anyway.

Framing the Debate

Example: The warmists know that they can't convince everyone 100% on the idea of catastrophic man-made GW/CC. So, through their media, they will allow non-believers the option of believing that man-made GW/CC is real, but that the long term effects are not fully known.

Why is this fallacious? This artificial framing of the debate sets the skeptic up for the question: *"Isn't it better to be safe than sorry?"* By leaving out the hoax option, the skeptic is thus manipulated into going along with the extreme schemes of the warmists --- just to be "safe."

Case Closed Argument

Example: Warmist claims that "the science is settled" or "the debate is over."

Why is this fallacious? Though it may sound impressive to the easily-impressed, this argument offers no facts or evidence. It's just meaningless rhetoric. When a matter becomes truly scientifically settled, there is no longer any need to declare that it is settled, for it becomes self-evident.

Declaring, with great passion, that "the debate is over" does not prove anything.

<u>Sound & Fury</u> (more of cognitive bias than a logical fallacy)

Over time, the sheer degree of power, hype and repetition behind GW/CC has convinced even the most "educated" people of the veracity of the "science." Due to the fact that no one likes to admit having been played for a fool -- and even fewer people can handle going against the opinions of "the majority" of their peers -- beliefs instilled by this type of surround-sound hype and propaganda can be *extremely* difficult to dispel. It is actually much easier to deceive people than to later on convince them that have been deceived.

Nonetheless, to a thinking man, pervasive hype, peer pressure and "public opinion" will never substitute for the totally non-existent evidence for catastrophic man-made GW/CC.

Messrs. Aristotle, Socrates and Plato are spinning in their graves over the fatal logical errors of the warmists and their dupes.

THE FIVE TYPES OF "SCIENTISTS" PROMOTING THE BAD LOGIC & SCIENCE OF MAN-MADE GW/CC

It is amusing to hear people speak of *all* "scientists" with such servile tones of deference and reverence – as if anyone with a science degree is some sort of hallowed, infallible and incorruptible demi-God or High Priest.

Of course, the *true* practitioners of *true* science merit our respect and gratitude. But make no mistake, like all other professions, the sciences are infested with two-bit pretenders who lack either the philosophical training or the ethical standards to be worthy of their titles.

The Fake Scientists promoting the hoax of GW/CC fall into one of five categories:

1) The Ambitious Climber

This sleazy character is very much aware of the money and notoriety that awaits the scientist who is willing to eagerly prostitute himself. The Ambitious Climber is a criminal who would pimp out his own mother for a shot at a Nobel Prize, or even just a puff-piece in the "prestigious" New York Slimes or Scientific American.

Of course, Ambitious Climber doesn't actually believe in the rubbish which he spews, but much like a fee-based "expert witness" in a court case, he will "prove" whatever his paymasters want him to "prove."

2) The Hungry Man

Not everyone with a science degree gets to go to work at an elite laboratory, university or NASA. Many a mediocre scientist has to "publish or perish," --- or, live off of research grants. Hungry Man operates in much the same way as Ambitious Climber. The difference is, he is probably doing it reluctantly just so that he can pay off his student loans or feed his family -- and not necessarily out of greed or for fame.

3) The Political Activist

A scientist with a Marxist-Globalist bent will promote GW/CC because its "solutions" involve the intense centralization of government power, more taxes, more control over private industry and property, and a diminution of the sovereignty of all nations. The Political Activist scientist doesn't believe in GW/CC either.

Some warmist scientists play the game for the money. Others are just plain mad men who want to be part of a Global dictatorship.

4) The Stupid Group-Thinker

Like a herd of stampeding lemmings, sheep or cattle; or a flock of birds sharply cutting left or right; or like teenage girls blindly following the latest fashion craze; our animal nature causes many of us to automatically gravitate toward the direction of our peer group. Scientists are by no means immune to this natural psychological phenomenon. Indeed, many of them, notwithstanding their proficiency with mathematics and computer-like recall of facts and figures, are actually among the stupidest and most obediently conformist semi-retards that you'll ever meet. Give one of these clowns a Quadratic Equation, and he'll solve it on the back of napkin. But ask "Rain Man" to tie his shoelace, and watch him freeze up in confusion.

Once a few "big names" and well-known publications begin leaning toward a belief in this or that "theory of the month," many of the lesser scientists will follow along. The Group-Thinker-Lemming differs from Ambitious Climber, Hungry Man, and Political Activist because he, drunk on math and easily influenced by "big name" scientists, actually *believes* in this crap!

Some of these types may actually be quite accomplished in their particular branch of science, but tend to defer to other "specialists" when they wander outside of their field of expertise. Once fooled by the slick sophistry and "theoretical physics" of his clever scientific superiors, the Group Thinker's own conceit and arrogance will render him impossible to ever deprogram with facts.

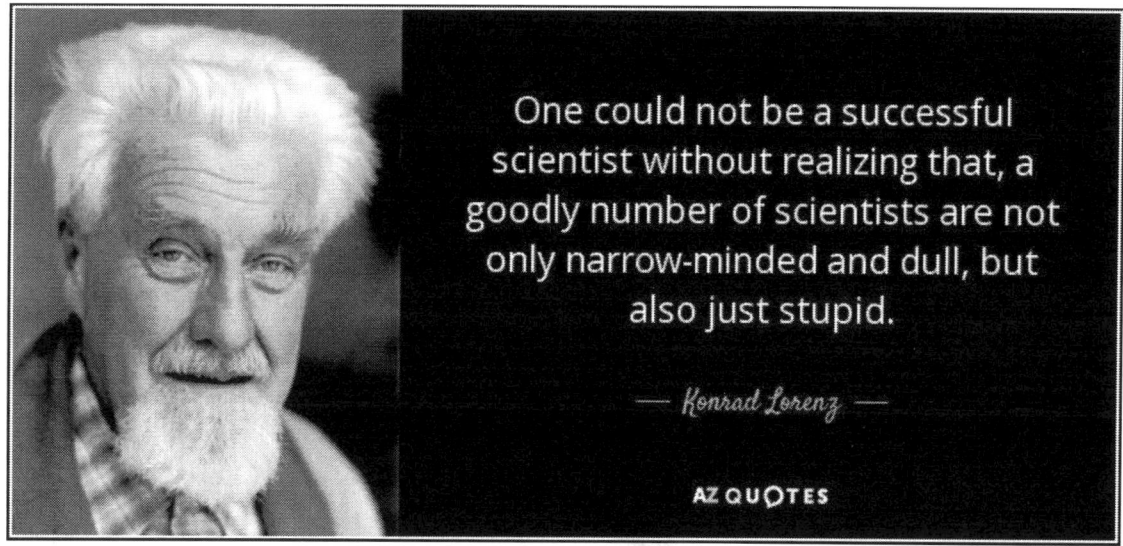

One could not be a successful scientist without realizing that, a goodly number of scientists are not only narrow-minded and dull, but also just stupid.

— Konrad Lorenz —

AZ QUOTES

5) The Coward

Partially out of concern for committing the "career suicide" which Dr. Judith Miller warned about, but mainly for fear of being labeled a "climate denier" by the bullying media and peers, the Coward will, at least meekly, go along with the hoax. Unlike the Hungry Man, many Cowards are financially stable. It is mainly peer pressure that motivates their support of the hoax. At most, the Coward might question "just how bad" GW/CC will be, but he will, all in all, join the bandwagon.

*

Put those five groups together, amplify their voices with 40 years of Fake News propaganda and "education" of school children -- and boom! You've got yourself a core group of controllable scientists who, with billions of dollars in cash and billions more in free media publicity behind them, can easily roll over the truth-telling scientists that the media strictly censors from its printed pages and airwaves.

That's how the social dynamic works. Criminals, communists, crackpots and cowards pretending to practice "science." Needless to say, greed, fame, fear, politics and group-think add up to lousy science – and even worse logic. By the way, this very same group dynamic can be applied to warmist journalists, politicians, professionals and "educators."

THE STUNNINGLY ILLOGICAL REASON GIVEN FOR THE EXTINCTION OF THE BRAMBLE CAY MELOMY

As of this writing, a Google Search for the term: "climate change endangered species" turns up an astonishing 7.2 million results. Sea turtles, penguins, ringed seals, certain species of possum, butterflies, birds -- the list is as varied as it is long.

Apart from the fact that the Earth is not warming, it is still ridiculous, *on its face*, to even suggest that the 1 degree or so of claimed increase could be threatening anything. We have already dispelled the myth of vanishing polar bears. The other tales of endangered species are just as unsubstantiated. We know that certain species are shrinking in number due to man-made factors such as excessive hunting or urban development, and those are indeed legitimate concerns. But do you really believe that penguins, who are capable of withstanding extreme cold as well as hot weather *(they can be found in most zoos!)* are going to become extinct over a 1 degree increase in *temps (a increase that is fictitious anyway).*

No false tale of extinction better illustrates the deliberate dishonesty and/or the flawed logic of the warmists than the "sad story" of the Bramble Cay melomys *(rats)* which disappeared from the face of the Earth, allegedly due to GW/CC. This could have been filed under the "Fake Science" section; but because the illogic of it all should be quickly evident to a child, and requires no other research at all, it's a better example of Fake Logic.

Have a look at just a few of the dramatic headlines, which were also picked up by TV News programs all over the U.S., Europe, Canada and Australia:

- **The Guardian / UK: *(June 13, 2016)* Revealed: first mammal species wiped out by human-induced climate change (3)**
- **The Guardian / UK *(June 29, 2016)*: 'Devastated': scientists too late to captive breed mammal lost to climate change (4)**

- **Environmentalist Blogger, Michelle Nijhuis** *(October 19, 2016)* **Who Killed the Bramble Cay Melomys (5)**
- **Washington Post** *(June 15, 2016)*: **The little Bramble Cay melomys is likely the first mammal claimed by man-made climate change, report says (6)**

To truly get a sense of the mendacious magnitude warmist illogic, read the following point-by-point, devastating and hilarious rebuttal to a "science" article which appeared in oh-so-"prestigious" the New York Slimes. This twisted logic, again, self-evident to a child, should forever destroy your faith in the intellectual supremacy of warmist "scientists" and journalists.

New York Times *(June 14, 2016)*

Australian Rodent Is First Mammal Made Extinct by Human-Driven Climate Change, Scientists Say (7)

Rebuttal (Edited) by

The Anti-New York Times

A pay-to-view feature of TomatoBubble.com, by M S King

Slimes: Australian researchers say rising sea levels have wiped out a rodent that lived on a tiny outcrop in the Great Barrier Reef, in what they say is the first documented extinction of a mammal species due to human-caused climate change.

Rebuttal: The first indicator of pseudo-scientific bull-shine here is the term "tiny outcrop." In science, one can only draw logical deductions from adequate random samples. For example: if a species of rodent had disappeared from the *whole* continent of Australia, one could infer that something weird had happened. But a "tiny outcrop" -- well, that sounds like anything could have happened to the rodents.

Slimes: The rodent was known to have lived only on Bramble Cay, a minuscule atoll in the northeast Torres Strait, between the Cape York Peninsula in the Australian state of Queensland and the southern shores of Papua New Guinea.

Rebuttal: So, Bramble Cay is "minuscule?" That sounds even smaller than "tiny outcrop." Just how "minuscule" is Bramble Cay? Well, it turns out that this isolated *(as in zero human population)* living laboratory which these "scientists" are basing their "Climate Change" extinction theory on is only about the size of a football field!

Slimes: The long-tailed, whiskered creature, called the Bramble Cay Melomys, was considered the only mammal endemic to the Great Barrier Reef.

Rebuttal: "The only mammal?" That's odd. How the heck did these rats even get there? Well, it is believed the rats first arrived by artificial circumstances -- like a shipwreck, visiting sailors or even by driftwood. Once on the island, the population struggled and did the best it could with the very limited local food supply.

Slimes: *"The key factor responsible for the death of the Bramble Cay melomys is almost certainly high tides and surging seawater, which has traveled inland across the island,"* **Luke Leung**, a "scientist" from the University of Queensland who was an author of a report on the species' apparent disappearance, said by telephone. *"The seawater has destroyed the animal's habitat and food source. This is the first documented extinction of a mammal because of climate change,"* he said.

Imported rats were never intended to thrive on an isolated sandy atoll the size of a football field. But Professor Leung is blaming Climate Change!

Rebuttal: We dug this up from the **Australian Department of the Environment (2008)**:

"With a population of less than 100 individuals inhabiting a single small sand cay whose existence is threatened by erosion, the Bramble Cay melomys is one of the most threatened mammals in Australia.

*The small population size and **the naturally unstable nature of Bramble Cay** has led to the species being listed as 'Endangered' under the Commonwealth Environment Protection and Biodiversity Conservation Act 1999 (EPBC Act) and 'Endangered' under the Queensland Nature Conservation Act 1992 (NCA).*

THREAT SUMMARY:

Exotic Predators: *The introduction of exotic predators or weeds to the cay could potentially be catastrophic, given the small and vulnerable nature of the melomys population.*

Isolation: *The cay's isolation, close proximity to Papua New Guinea and its use as an anchorage by fishing boats means there is a threat of pest and/or disease establishment. Two weed species are already present.*

Inbreeding: *Genetic analysis of this species reveals a level of inbreeding which theoretically could lead to inbreeding depression and ultimately extinction.*

You see folks, all manner of rodents rely upon explosive fertility in order for their species to survive -- *(hence the term, "---- like bunny rabbits.")*. But because our little rat friend near "the land down under" doesn't belong on tiny, isolated Bramble Cay in the first place, it simply could not establish itself in the long run. Pests, newly introduced predators, limited food supply and inbreeding evidently did their tiny population in.

That's basic elementary school science – *simple logic* -- not the commie crap these 'bought & paid for' diploma-decorated egg-heads are selling us.

FACT:
Rats multiply so quickly that two rats can have over a million descendants in 18 months.

The basic survival mechanism of any rat is to overpower predation and adversity through exponential reproduction. That never happened on desolate and food-scarce Bramble Cay.

Slimes: Anthony D. Barnosky, a professor at the University of California, Berkeley who is a leading expert on climate change's effects on the natural world,

Rebuttal: "Leading expert?" Says who? This is a typically pathetic example of **Argumentum ad verecundiam** -- *(appeal to authority or reverence)* a logical fallacy in which the case is made not by evidence, but by appealing to the alleged authority of some "expert."

Slimes: called the disappearance of the melomys "a cogent example of how climate change provides the coup de grâce to already critically endangered species."

Rebuttal: No, Barnosky. It's actually "a cogent example" of how an artificially imported species that had no business on that isolated sandy atoll faced insurmountable challenges. But you already know that.

Lectures --- books --- Show me the money, Barnosky.

Slimes: In August and September 2014, the scientists used traps and cameras to try to determine how many melomys were left, and they found none. No tracks were seen, and no scat was discovered.

Rebuttal: The "scat" is in the research itself.

Slimes: Scientists are not sure how the animals first arrived at Bramble Cay, but they theorize that they may have floated there on driftwood or arrived in sailing vessels.

"They may have been marooned on high ground at Bramble Cay," Dr. Leung said.

Rebuttal: In other words, the inbreeding rats had about as much business being on that tiny isolated atoll as an elephant in the Arctic. This then is the great "extinction" caused by "Climate Change!"

Slimes: *"And they were thriving there for a long time, but now they are gone."*

Rebuttal: "Thriving?" -- Thousands of miles away from home? On an isolated atoll the size of a football field, in the middle of an ocean, and with minimal food supply?

If loony Leung wants to find some rat scat, he need look no further than within the pages his own research paper. Shame on Queensland University for allowing this fraud to soil its name!

Slimes: Dr. Barnosky, who was not involved in the research, said the claim seemed "right on target to me."

Rebuttal: Wait a second! If Barnosky was "not involved in the research," then how is he in a position to evaluate the claim so quickly? Answer: Because this is all just one big you-scratch-my-back-and-I'll-scratch-yours game of international fraud!

Slimes: Darren Grover, a spokesman for WWF-Australia, said that a toad in Costa Rica was thought to have been lost to human-caused climate change, but that the melomys appeared to be the first mammal.

Rebuttal: "Thought to have been" -- "appeared to be." This is the minced illogical language of slick sophistry and deceiving --- not science.

Slimes: "Sadly, it won't be the last," he said.

Rebuttal: Oh the bloody drama!

The true logical conclusion and "environmental" lesson of the "extinction" of the Bramble Bay mouse? British rats don't belong stranded on an isolated Pacific atoll. Duh!

It doesn't take a Greek philosopher to figure out that dirty warmist rats are using illogical "extinction" propaganda to sell us their lies.

WARMIST ILLOGIC BLAMES OPPOSITE EVENTS ON THE SAME GW/CC

When logic and reasoned discourse fail, a bit of well-deserved and fact-based mockery is the best way to expose a lying con man, or just a well-meaning idiot. Have a good laugh at these goofy contradictory allegations -- with supporting headlines based on claims by grant-funded "scientists."

RAINFALL

GW/CC causes dry weather …

- **National Wildlife Federation: Global Warming and Drought (9)**

… and GW/CC *also* causes rainy weather…

- **The Guardian (UK): Global Warming is Increasing Rainfall Rates (10)**

TEMPERATURES

GW/CC causes hot weather…

- **National Wildlife Federation: Global Warming and Heat Waves (11)**

... and GW/CC *also* causes cold weather...

- **Scientific American: Global Warming Can Mean Harsher Winter Weather (12)**

POLAR ICE

GW/CC causes polar ice to melt ...

- **Weather Underground: Arctic Sea Ice Decline (13)**

... and GW/CC *also* causes polar ice to expand ...

- **Nature: Ocean Warming May Be a Major Driver of Sea-Ice Expansion in the Antarctic (14)**

WINDS

GW/CC causes weaker winds .....

- **Scientific American: Climate Change May Mean Slower Winds (15)**
- **Live Science: Global Warming Weakens Trade Winds (16)**

and GW/CC *also* causes stronger winds...

- **ABC (Australia): Global Warming: Australian scientists say strong winds in Pacific Behind pause in Rising Temperatures (17)**

- **University of California at Santa Cruz: Stronger coastal winds due to climate change may have far-reaching effects (18)**

GLACIERS

GW/CC causes glaciers to melt ...

- **National Geographic: The Big Thaw: As climate warms, how much, and how quickly, will Earth's glaciers melt? (19)**

and GW/CC *also* causes glaciers to expand …

- **National Geographic: Some Glaciers Growing Due to Climate Change, Study Suggests (20)**

ATOLLS

GW/CC causes low-lying atolls to disappear ...

- **United Nations University: Atoll islands and climate change: disappearing States? (21)**

and GW/CC *also* causes atolls to reappear …

- **Telegraph (UK): Pacific Islands Growing, Not Shrinking Due to Climate Change (22)**

HURRICANES / TYPHOONS

GW/CC causes more hurricanes ...

- **UK Independent: Global Warming is Causing More Hurricanes (23)**

And GW/CC *also* causes fewer hurricanes.

- **NASA: In a Warming World, the Storms May Be Fewer But Stronger (24)**

You see, with warmist fake logic, no matter what happens, the warmist "scientists" and their co-conspirators in the complicit media have an explanation! It is heads, they win, and tails, they win.

If you really want to enjoy a good laugh, have a look at the complete list of the side-effects caused by GW/CC -- meticulously compiled and linked at **www.numberwatch.co.uk/warmlist.htm**. **(25)** You'll find everything from bird strikes, to indigestion, to West Nile virus, to depression, to homelessness, to AIDS, to landslides, to car accidents and hundreds more hilarious "consequences." And you can be sure that for each of these "studies," the "scientists" behind them received a generous grant from some government or Leftist "philanthropic" foundation.

1. *Time Magazine: Too hot, it's GW/CC.* **2.** *Newsweek Magazine: Blizzards, that's GW/CC too. ---- All we can do is laugh at them.*

A FEW ILLOGICAL IRONIES OF "GREEN ENERGY"

CO_2 AND WARMTH CAUSE GREENERY

Ask any tree, bush or blade of grass what two things he would want for Christmas, most of all. Apart from adequate rainfall, our green friends would certainly respond: a) More CO_2 and b) more warmth.

So if "green energy" alternatives such as solar and wind power replace the CO_2 *(plant food)* emissions of gas, oil and coal, wouldn't we be *reducing* the amount of greenery? And if that CO_2 reduction, as the warmists claim, reduces the alleged warmth, wouldn't that reduce greenery as well? It would actually be more accurate to refer to the burning of "fossil fuels" as the true "green energy.

Newsflash, greenies! CO₂ is great for plants!

CO₂ LEVELS ARE HIGHER OVER JUNGLES AND FORESTS

Satellite observations show the highest levels of CO_2 are present over non-industrialized regions, such as the Amazon, not over industrialized regions. **(26)** Warmists have attempted to dispel this reality by claiming, without solid evidence, that the Amazon absorbs at least as much CO_2 as it emits, but the fact remains that the concentrations there are higher there than near industrial areas.

Warmists are always whining about "deforestation." But if undeveloped areas actually have higher CO_2 levels than developed areas *(probably due to decomposing vegetation, animal respiration, termite emissions etc)*, might we not be able to prevent the "melting of Antarctica" by burning out the whole Amazon and paving it? The one-time arson might spike CO_2 levels, but in the long run the CO_2 levels would drop and "save the planet." Right?

(* Because dogmatic warmist / liberals rend to be humorless and often lacking in wit, we wish to make clear that we are <u>not</u> actually advocating for the burn-down of the Amazon!)

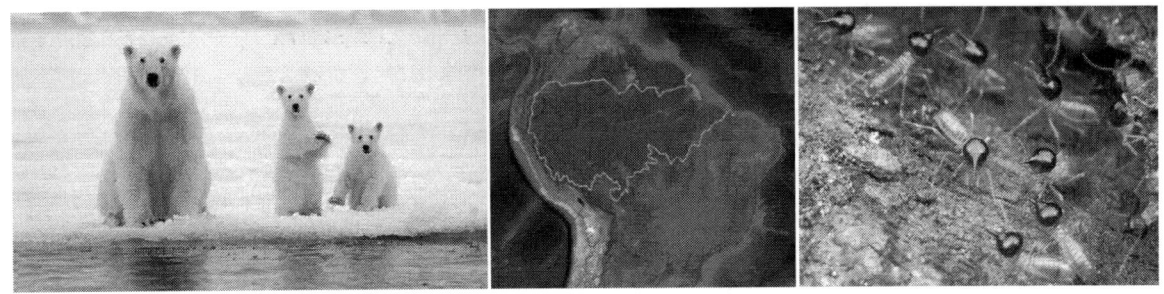

1 & 2: *Save the polar bears! Pave the Amazon!* **3:** *Amazon's large termites and earthworms are part of the reason why the jungle emits so much CO2.*

"GREEN ENERGY" SLAUGHTERS MILLIONS OF BIRDS

Warmists are always dishonestly whining about this or that "endangered species" being threatened by GW/CC. And oh how their fake tears flow when a bird or seal gets stuck in an oil spill! But the ongoing Bird-o-caust caused by their beloved *(and ugly as hell)* windmills doesn't seem to trouble them at all.

The mad drive to replace harmless coal with windmill farms is actually having a devastating impact on wildlife. With special government permission, as many as 3,000 Bald Eagles, America's symbol, are being ripped apart by wind turbines each year. If any American were to kill a single one of these magnificent and rare birds *(only 100,000 estimated in U.S.)*, he would be facing jail time! Evidently, the Green Mafia only cares about wildlife and "delicate ecosystems" when it suits their broader agenda.

The **Wall Street Journal** reported:

*"A 2013 study in the Wildlife Society Bulletin estimated that wind turbines killed about **888,000 bats and 573,000 birds (including 83,000 raptors) in 2012 alone.** But wind capacity has since increased by about 24%, and it could triple by 2030 under the White House's Clean Power Plan. "We don't really know how many birds are being killed now by wind turbines because the wind industry doesn't have to report the data," says Michael Hutchins of the American Bird Conservancy. "It's considered a trade secret .... golden eagles, which are rarer than bald eagles and are being whacked by wind turbines in far greater numbers."* **(27)**

So there you have it. The holier-than-thou warmists are actually anti-tree and anti-bird--- as well as anti-science and anti-freedom.

 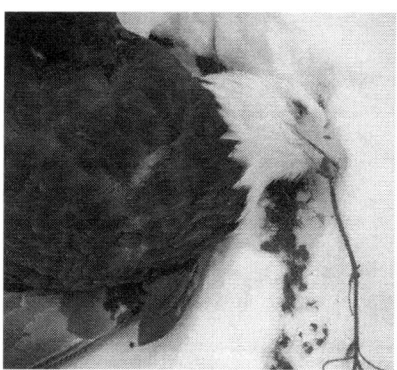 

A sin against God and Nature, all on the basis of a HOAX! Thousands of majestic eagles are being "wind-o-causted" -- a tragically poetic touch symbolizing the killing of America -- as coal miners lose their livelihoods.

BELIEFS INSTILLED BY FEAR AND INTIMIDATION SHUT DOWN THE BRAIN'S LOGICAL FUNCTIONS

It is an observable and proven fact of medical science that a state of fear, or even just low-level anxiety, activates the reactive centers of the human brain while "shutting down" the area in which thoughtful analysis takes place. Who among us has not made a bad decision while in a state of stress?

Has fear, social intimidation and even hysteria been used to shut down logical thinking and thus sell the public -- *especially women and children* -- on the hoax of man-made GW/CC? No words necessary to prove this point. Each picture is worth a 1000 words --- and there are *1000's* more images where these came from.

142

WHY DO WARMISTS AND FAKE SCIENTISTS HATE PHILOSOPHY SO MUCH?

The timeless principles of Graeco-Roman logic and philosophy that we have just reviewed *(not to be confused with the pointless mental masturbation that many modern pseudo-intellectuals like to engage in over a Starbucks' latte)* will expose a charlatan or an ignoramus every single time. Indeed, armed with nothing else but the earlier segment on Logical Fallacies, any "average Joe" can easily take down one of these "scientists" and humiliate him.

The Fake Scientist is thus compelled to declare: *"there is no debate"* because he knows that he cannot win a debate! Just like Count Dracula before a crucifix or sprinkled with Holy Water, the Fake Scientist will recoil in agony when confronted with the **Socratic Method** of inquiry.

Therefore, it is not surprising, but to be expected, that warmist Fake Scientists would express such open contempt and hostility for the millenniums-old discipline that should serve as the foundation of *all* intellectual pursuits and even common, every day understanding of life situations. Here they are, in their own words, mocking Philosophy – the very rules of thinking that we use to pursue truth and unmask lies and errors.

Neil DeGrasse Tyson: *"Philosophy is not a productive contributor to our understanding of the natural world .... It (philosophy) can really mess you up."* **(28)**

Stephen Hawking: *"Philosophy is dead. Philosophers have not kept up with modern developments in science. Scientists have become the bearers of the torch of discovery in our quest for knowledge."* **(29)**

Bill Nye: *"Philosophy is important for a while.... But you can start arguing in a circle.... Keep in mind, humans made up philosophy too."* **(30)**

You see, when a Fake Scientist needs to work around the eternal rules of lie-detection, he simply ridicules them, declares them "dead" or dismisses truth itself as "relative." It's sort of like a local burglar telling you that installing a home alarm system, adopting a big guard dog and keeping a loaded pistol under your bed aren't effective anymore; or a nervous criminal, under interrogation, insisting that polygraph *(lie detector)* tests are never accurate.

But philosophy is not dead. The theory of man-made GW/CC is, and Messrs. Tyson, Hawking and Nye all know it – which is why they want to replace philosophy *(Greek for "love of wisdom")* with "theoretical science" *(love of slick talking sophistry, rigged math equations and rigged computer models)*.

*Tyson, Nye and the "talking" stiff from "Weekend at Bernie's" all rely heavily upon classic logical fallacies to sell their nonsense. The only thing that can stop them is sound logic, aka **philosophy**.*

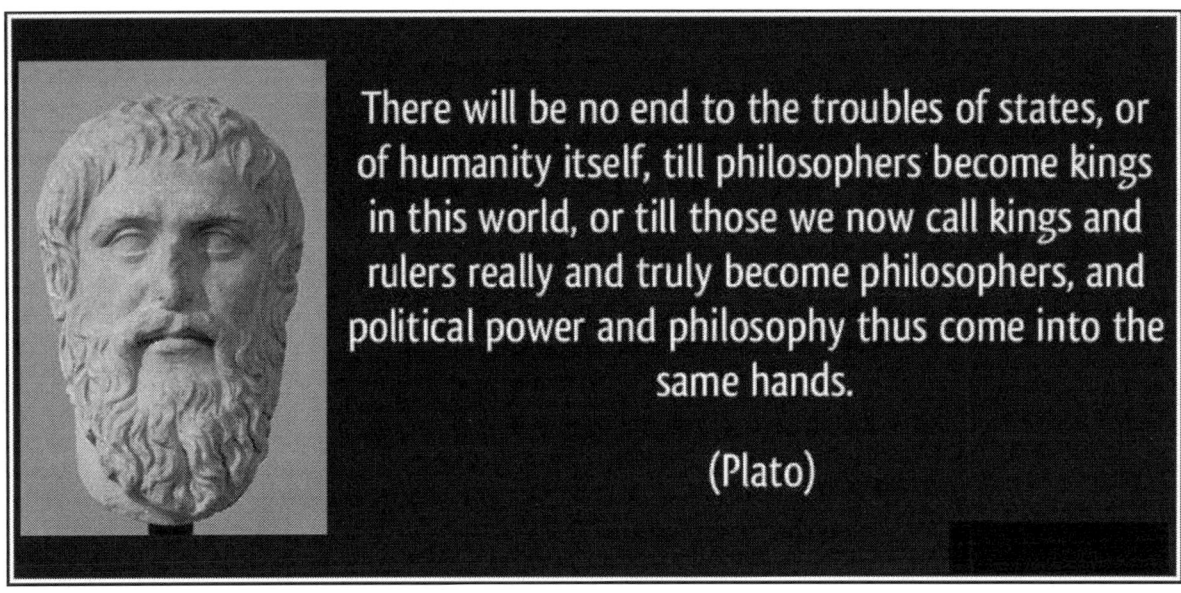

SECTION 4

THE FAKE ECONOMICS OF GLOBAL WARMING / CLIMATE CHANGE

"Under my plan of a cap and trade system, electricity rates would necessarily skyrocket. Regardless of what I say about whether coal is good or bad, because I'm capping greenhouse gases. Coal power plants, natural gas, whatever the industry --they would have to retrofit their operations. That will cost money and they will pass that money onto consumers." **(1)**

Barack Obama - 2008

147

COMMENTARY

Though it is not the *main* reason for the GW/CC hoax, the fact that some of the world's biggest money junkies are already profiting and stand to profit even more from "the green economy" adds enormous power and momentum to the hoax. Billionaires such as **Richard Branson, Warren Buffett, Jeff Bezos, Jack Ma, Bill Gates, Elon Musk, Tim Cook and Mark Zuckerberg** are just a few of the big names that are "all in" on "saving the planet."

One investment initiative, **Breakthrough Energy Ventures**, is headed by Bill Gates of Microsoft. The fund's backers are collectively worth about **$150 Billion**. This Billionaire Boys Club joined forces to form a coalition in support of the "clean energy" mandates expected to grow out of the 2015 Paris climate talks. Big money investors are drawn to green energy because the ever-increasing government force behind "saving the planet" is already creating a massive artificial market while guaranteeing profits with direct and indirect taxpayer subsidies.

Understanding the direction of the political winds, the oh-so- "civic minded" members of the Billionaire Boys Club have made their "smart money" bets on wind, solar, bio-fuels etc, and they don't like to lose! They will do whatever it takes to help rig the money game in their favor, including the funding of more Fake Science while publicly denouncing real scientists as "deniers." And that is why we witness the seemingly contradictory paradox of so many big "capitalists" in bed with big Globalist-Marxists on this fake "crisis" of GW/CC.

A level below the Billionaire Boys Club we have the Multi-Millionaires Boys Club – ambitious climbers who themselves hope to one day join the Forbes List ranks of the Big B Club. Taking their cue from the "smart money" crowd, many of the MM's have also gone "green." Understand that these self-important money and publicity junkies don't care about "saving the planet." Most of them are intelligent enough and well-connected enough to understand that the "science" behind GW/CC is fake.

THE WIND POWER SCAM

A reader named "Mark" who lives in the farming area of Iowa wrote to us at **TomatoBubble.com**:

"We have one of these windmills about 1/2 mile from our house. The small town I live near spent $450,000 to erect it and it only runs maybe 165 days a year. They've had to repair it 8 times since 2009 when it was installed.

I absolutely hate them. They are subsidized by us and who will take them down out of the corn fields when they stop working? They don't last long enough to remake the electricity sunk into their manufacture. Yet the government pays out about $11,000 per year for each one placed on your farm. Many days the wind doesn't even blow. They are also dangerous for planes and birds, and sometimes a giant turbine will bust off. Eventually, somebody is going to get hurt. I hate them! Total waste in so many ways." **(2)**

The poor beautiful eagles hate them too, Mark. Mark's story may be anecdotal, but in this case, his personal observations are indeed representative of the broader failure of wind power.

Both in terms of cheap energy and mass numbers of "green jobs," wind power has failed, in spectacular fashion, to deliver what the greenies and the big investors promised. The inefficient wind-power industry is very expensive, passing its costs on to consumers and taxpayers while creating relatively few jobs.

Ironically, wind farms must still rely upon the burning of fossil fuels when the wind does not blow and the turbines fail to spin. All throughout Europe, domestic coal stations have been closed in order to meet carbon-emission reduction targets. So when the wind farms come up short, nations must *import* coal and oil.

Apart from the fact that CO_2 emissions are harmless and that GW/CC is a hoax, wind farms have neither reduced emissions nor the cost of electricity. On windless days, large customers may be shut off temporarily. While the poor and the elderly have been hardest hit by high electricity bills, the only beneficiaries of the wind scam are the wealthy investors receiving government welfare.

Not only are the mechanical monsters expensive, ineffective, ugly and deadly, many have gone idle are costly to repair.

Bob Adelman writes in the **New American**:

Headline: 14,000 Idle Wind Turbines a Testament to Failed Energy Policies

"When Element Power announced on April 10 the closing of a deal to build wind turbines for Blackrock in Ireland, nothing was said about the more than 14,000 other wind turbines lying idle around the world. Instead, Jim Barry, managing director for BlackRock, the world's largest asset manager, expressed great pleasure at its new venture with Element.

....

Those 14,000 wind turbines lying idle in California's Altamont Pass, Tehachapin, and San Gorgonio areas and elsewhere around the world are testimony to the continuing and accelerating failure of hope over experience, funded with taxpayer monies. And these areas were selected as being "in the best wind spots on earth," which are now, according to Natural News writer Jonathan Benson, just "spinning, post-industrial junk which generates nothing but bird kills."

Once those taxpayer funds are withdrawn, the real economics of maintaining these expensive monstrosities are so overpoweringly negative that they are left to rot — skeletons proving the fraud and deceit of the whole global warming meme." **(3)**

He's right, you know.

The warmists can try to "fact-check" away these "inconvenient truths" all they want, but the reality remains: ***Wind farms do not generate electricity on a commercially viable basis.*** The fake industry survives only because of generous government subsidies, stolen from taxpayers and redistributed to wealthy investors, manufacturers and rent-seeking land owners.

It's an economic hoax, built upon a scientific hoax. In fact, it's so bad that even the mainstream warmist media and wind industry publications have had to concede certain "inconvenient truths."

Newsweek: *(April 11, 2015)*: **What is the true cost of wind power?**

"As consumers, we pay for electricity twice: once through our monthly electricity bill and a second time through taxes that finance massive subsidies for inefficient wind and other energy producers.

Most cost estimates for wind power disregard the heavy burden of these subsidies on US taxpayers. But if Americans realized the full cost of generating energy from wind power, they would be less willing to foot the bill – because it's more than most people think.

Over the past 35 years, wind energy – which supplied just 4.4% of US electricity in 2014 – has received US$30 billion in federal subsidies and grants. These subsidies shield people from the uncomfortable truth of just how much wind power actually costs and transfer money from average taxpayers to wealthy wind farm owners, many of which are units of foreign companies.

Financial advisory firm Lazard puts the cost of generating a megawatt-hour of electricity from wind at a range of $37 to $81. In reality, the true price tag is significantly higher.

This represents a waste of resources that could be better spent by taxpayers themselves. Even the supposed environmental gains of relying more on wind power are dubious because of its unreliability – it doesn't always blow – meaning a stable backup power source must always be online to take over during periods of calm." **(4)**

Wind Power Monthly: *(May 14, 2015)* **Annual blade failures estimated at around 3,800**

"Bellamy, co-founder of the renewables advisory firm Aarufield Ltd, pointed out that blade failures are the primary cause of insurance claims in the US onshore market. They account for over 40% of claims, ahead of gearboxes (35%) and generators (10%).

The wind industry also faces a struggle to secure the carbon fiber materials it needs for lighter and stronger blade designs, warned Bellamy.

"There's growing competition for these materials from the automotive and aerospace industries," he said. "And they are willing, and able, to pay more than we are....

Recent examples of blade failures include a blade from a Vestas V90 3MW turbine that snapped on a wind farm in the north of Denmark last year. At the time, Vestas said the winds were not particularly high.

In another case last year, GE was forced to replace 33 blade on its turbines at a Michigan wind farm after a blade broke on the project.

Possibly the biggest blade issue was faced by Siemens in 2013 when it was forced to curtail around 700 turbines worldwide. This was caused by a bonding failure in its B53 blade." **(5)**

It only takes one snapping blade to hurt or kill someone, which is already driving up liability insurance costs. Add another nail to the coffin of wind energy – all on the basis of a hoax!

THE SOLAR POWER SCAM

Similar to their extravagant claims about wind power, the green energy Mafia often uses a manipulative term to impress the easily-impressed. That term is "electrical generation capacity."

The empty term is designed to make one believe that wind or solar power have a limitless ability to provide electricity efficiently, affordably and reliably. But when it comes to energy, "capacity" is nothing but a technical term meaning the maximum *momentary* ability to produce electricity. It is not related to the *consistent*, ability to produce electricity, which is the only thing that matters to our modern economy.

The conventional forms of energy are reliable – always available, that is. Coal, oil, gas, hydro, and nuclear power are fixed and known because their fuel sources are stored and controllable. But for solar and wind energy, whose fuel sources are intermittent, unpredictable, and often unavailable, a term such "electrical generation capacity" meaningless, and the greenies know it!

A field of solar panels may operate near capacity on a sunny summer day, but when clouds and cooler seasons come, that ability can disappear. And at nighttime, the ugly panels obviously have zero electrical generation ability. For the purposes of providing society with cheap, on-demand electricity, this is useless. Disregard the solar cheerleading of the "experts" and the media. The fake industry of solar power cannot survive without massive government subsidies – a fact which even warmists will concede, albeit reluctantly, while arguing for even more subsidies!

New Jersey neighborhoods are littered with these hideous, expensive and inefficient eyesores atop telephone poles --- all on the basis of a hoax!

In Germany, where solar is all the rage, Germans must still purchase enough real capacity from reliable sources to give everyone the electricity they need. Thus, solar and wind are unnecessary and even problematic since they add unpredictable, destabilizing electricity to the power grid. This wastefulness explains why Germans pay 3-4 times for electricity than Americans do.

Writing in **Forbes**, author and free-market energy advocate **Alex Epstein** cautions us:

The Myth of Wind And Solar 'Capacity' *(March 29, 2016)*

*"Every time you hear some claim about wind and solar capacity remember that since their reliable capacity is zero, more "capacity" means more dead weight and higher prices--until and unless someone can create independent solar or wind power plants with affordable mass storage. **The lack of one single such plant in the world illustrates how inefficient and convoluted such an arrangement would be.**"* **(6)** *(emphasis added)*

Despite massive government subsidies, solar power remains more expensive than electricity generated the old-fashioned way. The subsidies, stolen from taxpayers, include a 30% Federal Solar Investment Tax Credit which was extended to 2019 by Congress. Like wind, this fake "industry" can only survive through government coercion, and only benefits the well-connected investors and solar businessmen.

One of the most criminally shocking examples of the Fake Economics of solar energy is that of the 2009 **Solyndra Scandal**. Solyndra was a California solar panel company that received $535 million dollars in low interest government "stimulus" loans from Green Obongo's 2009 "stimulus" boondoggle. Just two years later, Solyndra collapsed -- sticking taxpayers for the loss!

This non-viable and dishonest company would never have qualified for such generous financing from a bank or venture capitalist, but because the new clown in the White House, who had promised to "lower sea levels," was so determined to throw money away on "green energy," Solyndra's poor fundamentals were dutifully ignored by Federal officials who were under pressure to throw money away.

Obama, accompanied by Solyndra CEO Chris Gronet, looks at a solar panel during a 2010 tour of Solyndra. By the end of the following year, Solyndra had collapsed, taking $550 million of our tax dollars down the toilet with it.

THE "BIO-FUELS" SCAM

Pond scum and other "bio-fuels" were supposed to also be "revolutionary" fuel sources that would reduce our "dependence on foreign oil" and replace "planet warming" fossil fuels. In a truly free market, this failed idea would have rightfully died a natural death many years ago. But yet again, due to government *(taxpayer)* life support, the bio-fuel "revolution" lingers on year after year after year.

Biofuels are derived from sources such as algae, vegetable oil, corn *(Ethanol)*, waste crop material, and manure – just to name a few. Just like wind and solar, the bio-fuel scam is also inefficient, expensive, and dependent upon forced commerce and subsidies. **Robert Bryce**, a senior fellow at the Manhattan Institute, breaks the madness down for us with real numbers.

The National Review *(January 25, 2016)*: The Biofuel Scam Is Worse than Solyndra

The latest example of biofuel foolishness came last Wednesday, when the U.S. Navy announced the deployment of what it's calling the "Great Green Fleet." On hand in San Diego for the event were Navy Secretary Ray Mabus and Agriculture Secretary Tom Vilsack, a former Iowa governor. The two officials flew out to sea in a helicopter so they could watch the guided-missile destroyer get refueled with a blend of diesel fuel and biofuel. The San Diego sideshow is just the latest example of how biofuel boosters have used the military to get their hands on taxpayer money. On Earth Day 2010, the Navy flew an F/A-18 using a mixture of conventional jet fuel and biofuel derived from camelina, a plant in the mustard family... The cost of that fuel: about $67 per gallon.

In 2012, the Navy paid $424 per gallon for biofuel derived from algae. That same year, in another much-hyped example of the Great Green Fleet, it spent about $27 per gallon for 450,000 gallons of biofuel. One of the companies that got a lucrative biofuel contract from the military was the San Francisco–based Solazyme Inc. According to the Congressional Research Service, in 2009, Solazyme got a $223,000 contract for 1,500 gallons of algae-based motor fuel. That works out to $149 per gallon.

... *Solazyme has also been a big donor to Democratic causes, giving some $300,000 to Democratic candidates and committees. The company has also donated between $100,000 and $250,000 to the Bill, Hillary, and Chelsea Clinton Foundation.*

The Solyndra fiasco involved a $535 million loan that was supposed to help the company produce solar panels. Instead, in 2011, the company went bust. But the Solyndra loan represents only about half of the more than $1 billion that the Department of Energy has provided to various companies to research and develop cellulosic biofuels. That money has been spent despite research showing that production of cellulosic biofuels likely results in carbon-dioxide emissions that are higher than those from conventional gasoline.

The DOE money is only part of the madness. Also last Wednesday, Reuters noted that the Navy has "awarded $210 million to help three firms build refineries to make biofuels using woody biomass, municipal waste, and used cooking grease."

The bottom line here is obvious: American voters and taxpayers have been had. *Despite decades of hype, as well as years of mandates and subsidies, biofuels have never made a significant dent in our need for oil. And given the repeated failures of various biofuel companies, it's unlikely they ever will.*

Today, ***ethanol distilleries are consuming about 40 percent of all domestic corn*** *output in order to produce fuel equivalent of about 600,000 barrels of oil per day. And it took roughly four decades of mandates and subsidies for the corn-ethanol industry to grow to that size. Let's compare that result with what has happened in the oil patch. Since 2006, thanks to the shale revolution, domestic oil production has increased by more than 3.6 million barrels per day. Thus, in just this past decade, the oil sector has increased production by six times the total output of every ethanol distillery in America.*

That increased oil production happened because privately owned companies risked billions of dollars, and in doing so they innovated in everything from drill bits to mud pumps. The result: dramatic decreases in oil and natural-gas prices, which are now saving American consumers billions of dollars per year. Despite these facts, ***the biofuel lobby continues to have its way in Washington and even,***

unfortunately, at the Pentagon. *Such is the utter foolishness of American energy policy.* **(7)**

1. Robert Bryce's scathing fact-based exposes of the ridiculous costs of "green energy" are occasionally featured on cable new shows, but NEVER in the pages of The New York Slimes or any of the main network news programs. 2. The "Great Green Fleet" powers up with bio-fuel that costs as much as $430 per gallon. 3. The fertile fields of Iowa are for growing FOOD – not making Ethanol to mix with gasoline.

THE "CAP & TRADE" CARBON CREDITS SCAM

"Under my plan of a cap and trade system, electricity rates would necessarily skyrocket. Regardless of what I say about whether coal is good or bad, because I'm capping greenhouse gases. Coal power plants, natural gas, whatever the industry --they would have to retrofit their operations. That will cost money and they will pass that money onto consumers." -- Obongo, 2008 **(8)**

Notwithstanding the lasting legacy damage of Obongo's eight-year-long wrecking ball of extreme spending, reckless debt accumulation and job-killing regulation; let us thank our lucky stars that the "Cap & Trade" scheme which he tried to sell never became a reality.

The two major economic proposals for controlling the harmless emissions of plant food were carbon taxing and carbon trading. At the moment, the carbon trading scheme seems to be the preferred option. Carbon trading involves the establishment of a massive commission-based brokerage system in which a "Carbon Exchange" trades emissions under cap-and-trade schemes or with credits that pay for or offset greenhouse gases.

The scheme's governing body begins by setting a **cap** on allowable CO_2 emissions. It then distributes or auctions off emissions allowances that total the cap. Member firms that do not have enough allowances to cover their emissions must either make reductions or buy another firm's spare credits. Members with extra allowances can sell them or save them for future use.

In 2003, the **Chicago Climate Exchange** (CCX) was set up to be North America's first voluntary but legally binding greenhouse gas reduction and trading system for emission sources. The companies joining the exchange committed to reducing their aggregate emissions by 6% by 2010. The scheme accumulated and sold the non-existent "product" of "carbon credits." Though voluntary at first, in due time, companies worldwide would be forced into such exchanges. The profits to the new brokerage houses would be enormous! Obongo was correct when he proudly boasted:

"Under my plan of a cap and trade system, electricity rates would necessarily skyrocket." **(9)**

CCX ceased trading carbon credits at the end of 2010 due to inactivity in the American carbon markets. Unlike the solar, wind and bio-fuels scam, the idea of massive subsidies to the CCX, and/or a direct government mandate forcing thousands of businesses to join "Cap & Trade," were schemes which the public and even Obongo's Democrat friends in Congress were still not ready to swallow. But the idea isn't dead, just dormant and awaiting the day when the public can be "mobilized."

Obongo defines the "problem" of getting some of these schemes implemented:

"The problem is, can you get the American people to say that this (climate change) really important .... That requires mobilizing a citizenry, and getting them to understand what is at stake." **(10)**

In Europe, the intrusive European Union created its own cap-and-trade scheme in 2005. Like night follows day, the EU's "Cap-and-Trade" resulted in residential electricity prices "skyrocketing" **(11)**

The drastically increasing energy prices resulting from "Cap & Trade" schemes cause people to use less energy, businesses to hire less workers, and families to struggle with less disposable income. It is also damaging to human health – all on the basis of Fake Science!

But what do warmist politicians and their wealthy billionaire handlers care about "the little guy?" Screw the people! There is an easy fortune to be made in peddling the fake product of "carbon credits."

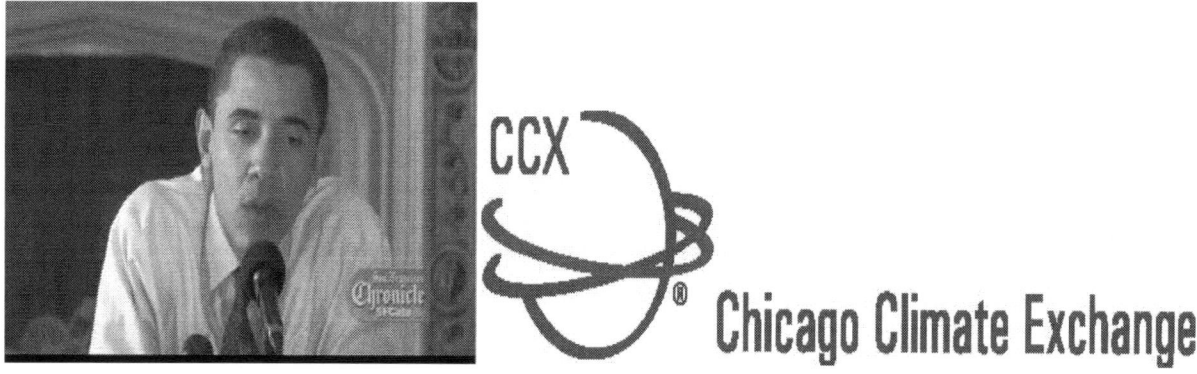

1. In cold blood and without batting an eyelash, Candidate Obongo openly told The San Francisco Chronicle that his plans would bankrupt coal companies. 2. The fraud that was the CCX collapsed because the warmists couldn't make membership and subsidies mandatory.

THE SMART METER SCAM

The ultimate objective of the control-freak governments pushing warmist nonsense is to control the public. Toward that end, under the pretext of "saving the planet," and under government pressure with generous government *(taxpayer)* grants, some utility companies in the US and EU have begun installing devices known as "smart meters" – which are supposed to help consumers track their peak usage and adjust their behavior to save energy.

A smart meter is an electronic device that records consumption of electricity in intervals of an hour or less and then relays that data back to the utility for monitoring and billing. Utilities raise rates is by instituting "time-of-day" rates— also called "peak pricing" or "time-of-use" rates—raising rates at peak times, but not always lowering them at other times. Every follow up analysis has shown that very few people have altered their energy consumption pattern both during peak and off-peak hours

"The report shows zero statistically different result compared to business as usual." -- **Susan Satter**, senior assistant Illinois attorney general for public utilities **(12)**

"(The smart grid) costs too much, and we're not sure what good it will do. We have looked at most of the elements of smart grid for 20 years and we have never been able to come up with estimates that make it pay." - **John Rowe**, CEO of Illinois utility ComEd, **(13)**

"No net economic benefit to ratepayers."
- **Bill Schuette,** Michigan Attorney General **(14)**

And Consumers Digest sums it up very neatly for us:

*"Smart-meter conversion represents little more than a boondoggle that is being foisted on consumers by the **politically influential companies that make the hardware and software** that are required for the smart-meter conversion. Smart meters are supposed to help to give you more control over your energy use. But many experts doubt that you'll ever see the electricity and cost savings that electric companies and smart-meter manufacturers tout."* (emphasis added) **(15)**

Apart from the fact that GW/CC is a hoax; if the smart meter / smart grid scam is such a failure, why keep the scheme in place? A statement from Cleveland Utilities gives us the answer:

*"The reason we're putting these meters in is to be able to **bill the time-of-use rates that are going to be mandatory**."* - **Cleveland Utilities** (emphasis added) **(16)**

We already know that the state *(NSA)* is monitoring our telephone calls, credit card purchases, and E-mails. One day, Big Brother will also be watching us naughty CO_2 "polluters," and punishing us accordingly when these meters send detailed information about how much electricity is being used at any given time. Frequent reports will allow the utility company to infer behavioral patterns for the occupants of a house, such as when the members of the household are probably asleep or away.

In Australia, debt collectors may access the data to know when debtors are at home. In Texas, a police agency secretly collected smart meter data from thousands of homes to determine which ones were using more energy than normal – the objective being to locate potential marijuana growers.

Smart meter data can reveal more than just how much power is being used. A sampling of power levels at very brief intervals can identify when certain electrical devices are in use and even what TV channel is being viewed. If the hoax of GW/CC is not widely exposed and killed, the day may come when this potentially intrusive technology will be used as a tool of **behavioral control**. That's what the high-level power planners behind the grand hoax have wanted all along. This real agenda will be explained in the final section of this book, the grand finale in which we will define both the "who" and the "why" behind all of this dangerously destructive criminal insanity.

Behind the green mask, is the totalitarian face of 'Big Brother."

THE HUMAN COST OF "SAVING THE PLANET"

The hoax of GW/CC would be very funny, were it not for hardships it is causing and the lives it is destroying. For *all of us* living in the demented western world where the new religion of "saving the planet" from GW/CC has led to the regulatory classification of harmless CO_2 as a "pollutant", electricity rates have continuously climbed upward as over-regulated energy providers pass the overheard costs onto consumers. **(17)** See if you can dig up an old bank statement or electricity bill from 10 or 15 years ago and note the big difference you are paying month over month. For lower income families, this diminution of disposable income makes life even harder.

Coal miners in Europe and America have suffered the most – with 83,000 American coal miners losing their job as a direct result of Obongo's EPA sanctions against CO_2. **(18)** In states like Kentucky, West Virginia, Ohio, Illinois and Pennsylvania, hard working men with families to feed have railroaded into abject poverty, all on the basis of a damn hoax!

The cost of "saving the planet" also drives up the cost of new cars (*due to stricter and unnecessary emissions requirements)*, and forces poor people out of their used cars because "repairs" for cars that fail the emissions inspection can be very expensive. How ironic that these warmist "liberals," who all profess to love the poor so much, are supporting a colossal fraud which transfers wealth from the poor and the middle to the billionaire green investors.

Killing the African Dream

Africa, though generally poor, is very rich in resources and has no lack of unemployed men just dying to work *(which is why so many are flooding into Europe!)* With a bit of outside expertise from the West and China, the continent has huge potential to develop agriculturally, industrially and economically.

But yet again, it is the very same "liberal" political forces who profess to love the poor and the Blacks so much that are keeping the Africa nations from lifting millions of their people out of extreme poverty. Corrupt African politicians, kept in on the Western payroll known as "foreign aid," help to facilitate this oppression.

Some *health clinics* in Africa are powered by solar panels that do not provide enough electricity for both the medical refrigerator and the lights at the same time. Kenyan author and free market economist **James Shikwati** explains how the warmists are keeping the "bruthas " in Africa down.

"There's somebody keen to kill the African dream. And the African dream is to develop. Renewable power is a luxurious experimentation ... I don't see how a solar panel is going to power a steel industry – rather a transistor radio. We are being told, 'Don't touch your resources. Don't touch your oil. Don't touch your coal.' That is suicide." **(19)**

Shikwati describes the idea of restricting the world's poorest people to alternative energy sources as *"the most morally repugnant aspect of the Global Warming campaign."* **(20)**

He's right, you know.

1. Few people understand how regulations linked to the warmist hoax are jacking-up their electricity bill 2. Unemployed coal miner Eddie Jones looks for jobs on a computer at the Kentucky Career Center 3. Kenyan economist James Shikwati doesn't want "foreign aid" for Africa. He wants development – but the western warmists say "No!"

SECTION 5

THE TRUE POLITICS OF GLOBAL WARMING / CLIMATE CHANGE

*"Tonight, I speak before you not as a candidate for President, but as proud citizen of the United States, and a fellow **citizen of the world**."*

*"This is the moment when we must come together to save this planet. Let us resolve that we will not leave our children a world where the oceans rise and famine spreads and terrible storms devastate our lands. Let us resolve that all nations – including my own – will act with the same seriousness of purpose as has your nation, and reduce the carbon we send into our atmosphere. This is the moment to give our children back their future. **This is the moment to stand as one**."* **(1)**

-Barack Obama, 2008, Berlin

How is it possible that a preposterous fairy tale, a monstrous mendacity, a stupendous superstition of such massive proportions as the GW/CC hoax has been able to suck in so many otherwise intelligent people into its vortex? A reasonable man has got to conclude that for this surreal and sad situation to have come about; there has to exist some sort of omnipotent criminal conspiracy.

Such a "Mafia" -- international in scope -- would have to encompass not only the political realm of social affairs, but also the major media organizations, academia, lower education, elements of big business, Hollywood and even parts of the clergy. Is there such a Globalist Mafia? If so, what are its true motives and final objectives?

It is beyond the scope of this book to adequately define and expose the who, what, when, how and why of this Globalist Mafia. For a full explanation, refer to the 2-volume set, *Planet Rothschild*, by yours truly, available at Amazon. But we certainly can, in this limited format, provide the reader with just enough of the "big picture" to understand why the GW/CC hoax has been pushed so hard for nearly 40 years – and with no end in sight.

As a member of an elite political family, **Edith Kermit Roosevelt** (1927-2003), whose famous relatives included Grandfather **Theodore Roosevelt**, and cousins **Franklin and Eleanor Roosevelt**, broke ranks with the Roosevelt Dynasty and tried to warn us about such a "Mafia." In an article titled, *Elite Clique Holds Power in U.S.,* Roosevelt wrote:

"The word 'Establishment' is a general term for the power elite in international finance, business, the professions and government, largely from the northeast, who wield most of the power regardless of who is in the White House. Most people are unaware of the existence of this 'legitimate Mafia.' Yet the power of the Establishment makes itself felt from the professor who seeks a foundation grant, to the candidate for a cabinet post or State Department job. It affects the nation's policies in almost every area." **(2)**

Edith Roosevelt wasn't the only person of high standing to tell of this Mafia. **Admiral Chester Ward**, an Advocate General of the U.S. Navy, was invited into membership into one of this Mafia's key front institutions – The New York City-based (*sister groups in Chicago and London*) **Council on Foreign Relations (CFR)**. Shocked by what he learned about "The Establishment" and its ultimate objectives, Ward co-authored a book in 1975, titled, *Kissinger on the Couch*, in which he wrote:

"Once the ruling members of the CFR have decided that the U.S. Government should adopt a particular policy, the very substantial research facilities of CFR are put to work to develop arguments, intellectual and emotional, to support the new policy, and to confound and discredit, intellectually and politically, any opposition." **(3)**

The GW/CC hoax had yet to be hatched at the time of Admiral Ward's book, but it should be noted that the CFR and its Quarterly journal, *Foreign Affairs*, have long promoted, aggressively so, the GW/CC scam and the "solutions" for it.

Through its publications and lectures, the CFR has been promoting the GW/CC hoax for nearly 40 years.

Roosevelt and Ward spoke out against this Global Mafia from a position of opposition. Just to balance out the sources of evidence, let us review another powerful testimony which comes to us from an elite Georgetown University history professor named **Carroll Quigley** *(1910-1977)* – a man who President **Bill Clinton,** during his acceptance speech at the 1992 Democrat Convention, publicly praised as one of his mentors. **(4)** From Quigley, who proudly wrote in *Tragedy and Hope* that he was close to the Establishment and *agreed with* its objectives, we learn the following:

*"There does exist, and has existed for a generation, an international Anglophile network which operates, to some extent, in the way the ... Right believes the Communists act. In fact, this network, which we may identify as the Round Table Groups (CFR**), has no aversion to cooperating with the Communists, or any other groups**, and frequently does so.*

*I know of the operations of this network because I have studied it for twenty years and was permitted for two years, in the early 1960's, to examine its papers and secret records. I have no aversion to it or to most of its aims and have, for much of my life, been close to it and to many of its instruments. I have objected, both in the past and recently, to a few of its policies ... but in general my chief difference of opinion is **that it wishes to remain unknown,** and I believe its role in history is significant enough to be known.* **(5)** *(emphasis added)*

THE GLOBALIST MOTIVE

Thank you, Ms. Roosevelt, Admiral Ward and Professor Quigley for sharing with us your intimate knowledge of this legalized Mafia. Is there anything you all can tell us about its true motives and long term objectives?

Edith Roosevelt:

"What is the Establishment's view-point? Through the Roosevelt, Truman, Eisenhower and Kennedy administrations its ideology is constant: That the best

*way to fight Communism is by a **One World** Socialist state governed by 'experts' like themselves. The result has been policies which favor the growth of the super state and the gradual surrender of United States sovereignty to the United Nations aggression."* **(6)**

What say you about the ultimate objective of the Globalists, Admiral Ward?

*"[the CFR has as a goal] submergence of U.S. sovereignty and national independence into an all-powerful **One-World** government.... this lust to surrender the sovereignty and independence of the United States is pervasive throughout most of the membership.... In the entire CFR lexicon, there is no term of revulsion carrying a meaning so deep as 'America First.'"* **(7)**

Professor Quigley, we know that unlike Ms. Roosevelt and Admiral Ward, you are a supporter of the Establishment. What do think about their allegations of a "one world" conspiracy?

*"The powers of financial capitalism had another far-reaching aim, nothing less than to create a **world system** **of financial control in private hands able to dominate the political system of each country and the economy of the world as a whole.** This system was to be controlled in a feudalist fashion by the central banks of the world acting in concert, by secret agreements arrived at in frequent private meetings and conferences."* **(8)**

Are you connecting the dots yet, dear reader? **The ultimate objective of the GW/CC hoax is to establish a set of global controls and taxes over energy usage that would effectively establish a "One World" totalitarian system** --- a global bee-hive with a tiny elite at the very top, and the rest of us tax-paying, perpetually-indebted, high-rise dwelling, mass-transit-riding, worker-bee "human resources" supporting their dreamed-of super structure referred to as "**The New World Order**."

The GW/CC hoax, along with various other international scams such as "free trade," the endless "War on Terror" and "regime changes" all serve the higher purpose of ultimately fusing predatory debt-based Monopoly Capitalism together with a modern form of political Communism to create a Frankenstein-hybrid global tyranny from which no nation, and no individual, can ever escape. It will be like an EU for the whole world. You see, it's never been about "saving the Planet." No. It's all about *enslaving* the Planet. Control the flow and distribution of energy *(the lifeblood of all modern economies)* and you control the world. It's that simple. Get it now?

*1. "The New World Order" by H. G. Wells (1940)   2. "One World" by 1940 presidential candidate, Wendell Wilkie (1943)   3. An opposing voice, "**Planet Rothschild**" by M S King (2016)*

QUOTES THAT PROVE A GLOBALIST CONSPIRACY

The ranting of a "conspiracy theorist," you say? If Roosevelt, Ward and Quigley, as high up the societal "food-chain" as they were, weren't enough to convince you that the world government movement is indeed real, then let's hear it from a few other well-connected individuals, both pro and con. These are all actual quotes – look em' up!

PRO WORLD GOVERNMENT

Strobe Talbott *(Deputy Secretary of State under Clinton)*: *"In the next century, nations as we know it will be obsolete; all states will recognize a **single, global authority**. National sovereignty wasn't such a great idea after all."* **(9)**

Walter Cronkite: *(legendary newsman, accepting the "Global Governance" award): "We Americans are going to have to yield up some of our sovereignty.* ***That's going to be, to many, a bitter pill…. Today we must develop federal structures on a global level.*** *We need a system of enforceable world law --**a democratic federal world government**…. Within the powers given to it in the Charter, the U.N. could then deal with matters of reliable financing, a standing U.N. peace force, development,* ***the environment****…. and human rights…. The only way we who believe in the vision of a democratic world federal government can effectively overcome this reactionary movement is to organize a strong educational counteroffensive stretching from the most publicly visible people in all fields to the humblest individuals in every community."* **(10)**

Hillary Clinton *(responding to Cronkite's above comments via video feed for the same event):* "Good evening and ***congratulations, Walter, on receiving the World Federalist Association's Global Governance Award.*** *For more than a generation in America, it wasn't the news until Walter Cronkite told us it was the news. Every night at 6 o'clock, we listened as you explained the complex events of the day. … You became a trusted member of my family and the families across America. For decades, you told us "the way it is," but tonight, we honor you for fighting for the way it could be. .. So thank you, Walter. Thank you for inspiring all of us to build a more peaceful and just world. We are still listening to your every word."* **(11)**

Albert Einstein: *"**A Federal organization of the nations of the world** is not only possible but an absolute necessity if the conditions on our planet are not to become unbearable for men."* **(12)**

Pope Francis: *"A global consensus is essential for confronting the deeper problems, which cannot be resolved by unilateral actions on the part of individual countries..... International negotiations cannot make significant progress due to positions taken by countries which place their national interests above the global common good.. .. Global regulatory norms are needed to impose obligations ... What is needed, in effect, is an agreement on systems of governance for the whole range of so-called "global commons... ... it is essential to devise stronger and more efficiently organized international institutions, with functionaries who are appointed by agreement 50 among national governments, and empowered to impose sanctions.... **there is urgent need of a true world political authority, ... One authoritative source of oversight and coordination is the law, which lays down rules for admissible conduct in the light of the common good.** ...Here, continuity is essential, because policies related to climate change and environmental protection cannot be altered with every change of government."* **(13)**

Barack Obama: *All nations must come together to build a stronger, **global regime**.* **(14)**

ANTI-WORLD GOVERNMENT

Senator Barry Goldwater: *"David Rockefeller's newest international cabal, the Trilateral Commission.... is intended to be the vehicle for **multinational consolidation** of the commercial banking interests **by seizing control of the political government** of the United States. Perhaps its most important aspect is its commitment to facilitating high-level international dialogue."* **(15)**

172

Congressman Larry MacDonald: "*The drive of the Rockefellers and their allies is to create a one-world government, combining super-capitalism and Communism under the same tent all under their control. Do I mean conspiracy? Yes I do.*

I am convinced there is such a plot, international in scope, generations old in planning, and incredibly evil in intent" **(16)**

Congressman Ron Paul: "*We're only supposed to talk about internationalism, globalism, one-world government. To talk about the interests of the United States, in this city, is considered very negative.*" **(17)**

Russian President Vladimir Putin: "*The UniPolar world refers to a world in which there is one master, one sovereign, one center of authority, one center of force, one center of decision-making. This is pernicious - At its basis there can be no moral foundations for modern civilization.*" **(18)**

Now you know why the media and the US / EU political Establishment hate Putin so much!

President Donald Trump: "*We will no longer surrender this country or its people to the false song of Globalism.*" **(19)**

Trust me, there are plenty more where those came from. The Globalist movement is very real, and it's been building for many years with exclusive clubs and elite families at its self-perpetuating core. The **Rockefeller Family** has been at this game for about 120 years and counting; as has the **Sulzberger-Ochs Family** which has owned The New York Slimes since 1896. But that's nothing compared to the **Rothschild Family** – a internationalist banking-political dynasty began its ascent during the late 1700's!

THE "BOGEYMAN CARD"

Now that you understand *(at least partially)* the "who" and the "why" of the GW/CC hoax, let's talk a bit about the "how." That is, how did the Globalist mafia manage to scare so many people into believing this crap? Yes, we know that they control the media, the government, the schools, and even that Marxist Pope in the Vatican. And yet, all of that control wouldn't be enough to accomplish their geo-political objectives without a tactic. And that tactic is: FEAR – or what we can rightly describe as the playing of, "The Bogeyman Card."

Since time immemorial, governments of every stripe have, for higher purposes both noble and ignoble, played the Bogeyman Card to frighten and control the masses. It's the oldest trick in the book and it damn near works every time. Following are just five of many historical precedents.

The Setting: After promising the American people that the US would not enter the war in Europe, President **Woodrow Wilson**, under false pretenses, sought and obtained from the Congress a Declaration of War against Germany and Austria-Hungary. In order to build public support and encourage recruitment, what did Wilson and his chief propagandist, **Edward Bernays** *(the father of modern advertising)* do? They played the Bogeyman Card!

Overnight, both the Kaiser *(who did not start the war and had been proposing to end it all along)* and the German people were suddenly transformed into bloodthirsty "Huns" that had to be stopped from raping women and "crucifying babies" – accusations which even Establishment historians now admit were total fabrications. The selling of the bogeyman "Hun" served the purpose of maintaining support for Wilson's war to establish what he openly referred to as a **"new international order."** It also aided recruitment as brainwashed young men, sickened by the tales of women and babies being crucified by the Germans, rushed to volunteer for the Army or Navy.

"Destroy the Mad Brute" --- "They Crucify" --- "Beat Back the Hun" --- "Only the Navy Can Stop This"

*Wilson hired **Edward Bernays**, a nephew of Sigmund Freud and the author of the book "Propaganda," to play the Bogeyman Card against Germany.*

Bernays: "The conscious and intelligent manipulation of the organized habits and opinions of the masses is an important element in democratic society.... In almost every act of our daily lives, whether in the sphere of politics or business, in our social conduct or our ethical thinking, we are dominated by the relatively small number of persons... who understand the mental processes and social patterns of the masses. It is they who pull the wires which control the public mind." (20)

Bogeyman Card: Case Study #2 / 1940

Winston Churchill Uses "Gas" Scare Keeps World War 2 Going

The Setting: Germany had just pushed the invading British forces off of the continent at Dunkirk and made peace with France. Hitler ordered peace leaflets to be dropped over London, pleading for an end to the war. **(21)** Through third parties, the Germans attempted to make an honorable peace with Great Britain and withdraw from any occupied countries. This growing peace movement even reached the levels of Prime Minister **Winston Churchill**'s own War Cabinet. **(22)** But Churchill, knowing that **Franklin D Roosevelt** would eventually bring the

United States into the war, was committed to keep the fight going. So, what did Churchill do to control his people? He played the Bogeyman Card!

As Churchill delivered "inspirational" radio addresses, British propaganda warned of imminent poison gas attacks from the German Luftwaffe *(air force)*. Gas masks were handed out to the frightened public as any news of Hitler's peace offerings was blacked-out by the British press. Frightened and confused, the British peace movement fizzled out. The Bogeyman Card worked, and the war continued.

Churchill's fear-mongering and ridiculous poison-gas-attack propaganda thwarted the efforts of members of his own cabinet to at least hear what Hitler was proposing.

Bogeyman Card: Case Study # 3 / 1947-48

Truman Uses the Stalin Scare to Sell the Marshall Plan

The Setting: With post-war Europe in a desperate situation and ripe for effective U.S. takeover of its political systems, **The Council on Foreign Relations (CFR)** devised a massive package of financial aid which came to be known as "The Marshall Plan," after General George Marshall who was trotted out to announce it. Also known as the European Recovery Act, the $17 billion dollar scheme *(which will end up consisting of 85% grants or loan forgiveness and only 15% in paid back loans)* faced opposition from Congress and the American people. So, what did President Harry Truman do to control his people? He played the Bogeyman Card!

Rather than sell the Marshall Scam on humanitarian grounds, the Communist Bogeyman was used to scare the bill into law. In their book, The Wise Men, authors Walter Isaacson and Evan Thomas explained how Truman advisor and World Bank President **John McCloy** *(and others)* got the expensive plan to pass:

"'People sat up and listened when the soviet threat was mentioned.' It taught him (McCloy) a valuable lesson: One way to assure that a viewpoint gets noticed is to cast it in terms of resisting the spread of Communism." **(23)**

Meanwhile, in France, **Pierre Mendes-France**, the French executive director of the World Bank, made this candid observation:

"The communists are rendering a great service. Because we have a 'communist danger,' the Americans are making a tremendous effort to help us. We must keep up this indispensable communist scare." **(24)**

The Marshall Plan, with its embedded scheme to begin the process of forming the European Union, soon passed the anti-Communist Congress under false pretenses. The Bogeyman Card worked!

1- The only way to sell the Globalist Marshall Scam was to scare people with the prospect of a Stalinist takeover of Europe. 2- Many years before the EU was finalized, Marshall Plan propaganda posters were already selling the idea of a United Europe.

The Setting: Long before the first atomic bomb was constructed, 1914 to be precise, the famous author H G Wells, in his fictional book, ***The World Set Free***, put forth the idea of establishing a one-world government to save the world from the destruction of atomic weapons. When the Soviet Union detonated its first atomic bomb in 1949, this movement to hand over atomic weaponry to an all-powerful United Nations was no longer fictional.

Among the open supporters of this radical scheme were scientists like **Robert Oppenheimer** and **Albert Einstein**, and political heavy hitters such as President **Dwight D. Eisenhower** and his confidante-brother, **Milton Eisenhower**.

The Chicago Tribune *(February 10, 1952)*: **Brother Milton a Worry to Aids of Eisenhower**

*"In his UN service he (Milton Eisenhower) advocated a mighty UN military force as the sole means to peace, holding the force **'should be larger than that of any member state or likely combination of member states.'"** (emphasis added)* **(25)**

How was the Globalist President Eisenhower ever going to sell the radical **"Atoms for Peace"** proposal for total UN takeover of atomic power to the American people and the world? Eisenhower and his allies played The Bogeyman Card!

All throughout the 1950's, in collaboration with the media and academia, the people were terrorized by a **"Doomsday Clock"** put together by "concerned scientists." Working with the national teacher's union, children were terrorized by being forced to participate in bomb shelter drills and being made to watch **"Duck & Cover"** atomic-war survival tutorials in class. This silly scare was all phony, of course, because whatever one may say about the Soviet Union, its leaders were not suicidal and never had any intention of launching a first-strike against a superior-powered United States.

Although total military power was never handed over to the United Nations *(far too radical a concept for those days)* The "Doomsday Clock" and the "Duck & Cover" Bogeyman Cards did however succeed in promoting a closer political and economic union between the nations of "the free world."

Thinking that they could eventually work with him to build an integrated 'New World Order', the Globalists turned a blind eye to Stalin getting an atomic bomb --- and then they used the fear of the weapon to manipulate the world into more and more Globalism.

The goofy "scientists" behind the "Doomsday Clock" and "Bert the Turtle" of the "Duck and Cover" idiocy were intended to frighten the weak-minded into accepting Globalism as the "solution" --- EXACTLY like today's GW/CC hoax!

*

The Bogeyman Card works great with kids – adults too.

Bogeyman Card: Case Study #5 / 2000's

Bush Uses "Weapons of Mass Destruction" & Terrorism Scare

(We can combine these two Bogeymen Cards because they are somewhat related)

The Setting: President **George W. Bush** and his gang of "neo-conservative" warmongers had put together a plan to overthrow seven Middle Eastern & African governments within five years **(26)**. To kick off the propaganda campaign for the coming war, it was alleged that Iraqi leader **Saddam Hussein** was one of the plotters behind the terror attacks of 2001. This bare-faced lie would not fly. So, what did the Bush-Cheney gang do? They played the Bogeyman Card!

Day after day, week after week, for about one whole year, high ranking members of the Bush gang and their allied media mouthpieces scared the public with tall tales of Saddam's "Weapons of Mass Destruction" – aka WMDs" In spite of Iraq's denials and Saddam's open invitation for unlimited inspection, the lie of WMDs was repeated over and over again. The public was told that "mushroom clouds" over the Middle East were imminent if the U.S. did not go to war with Iraq.

"We cannot wait for the smoking gun in the form of a mushroom cloud," **(27)** declared the Bushies in unison. Finally, to close the sale, the oh-so-"esteemed" General **Colin Powell** was trotted out, scary nuclear-related props and all, to lie his ass off before the United Nations. The battle for public opinion was won. The

Bogeyman Card worked, as the never-ending wars and proxy wars in the region continue to drain US finances while bringing misery and death to millions of innocent people.

History, both ancient and recent, has demonstrated time and again that the Bogeyman Card works even better than the Sympathy Card and the Atrocity Card. The elite Globalist Mafia understands this. GW/CC is nothing but a Bogeyman Card for moving us closer to Global political and economic domination – a **New World Order** --- which is something that you really *should* be worried about!

*1. Colin Powell scared us into the Iraq war with his little test tube prop at the UN. 2. Time Magazine scared us with a terrorist time bomb on its cover. 3. The New York Daily News scared us with a **make-believe** ISIS "beheading."*

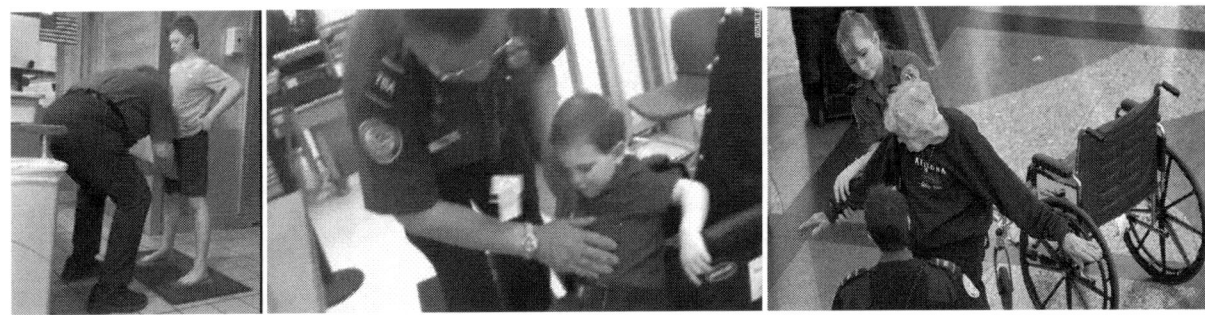

At US airports, even children and crippled old ladies are suspected of working for the Bogeyman.

The Kaiser Bogeyman is coming! The Hitler Bogeyman is coming! The Atom-Bomb Bogeyman is coming! The Arab Bogeyman is coming!

The Climate Bogeyman is coming!

THE TOTALITARIAN FACE OF THE WARMISTS AND THEIR NEW WORLD ORDER

Among the elite ranks of those working for a **New World Order** are many starry-eyed idealists who, believing that the new system will bring about heaven-on-earth, will rationalize away all of the trickery and deceit that the controllers are using to bring it about. These are the types that take to heart the words of **John Lennon**'s famous song, *Imagine*, -- the unofficial anthem of the Globalist movement.

The relevant excerpts:

*Imagine there's **no countries***
It isn't hard to do
Nothing to kill or die for
And no religion too
Imagine all the people living life in peace, you
You may say I'm a dreamer
But I'm not the only one
I hope someday you'll join us
And the world will be as one
Imagine no possessions
I wonder if you can
No need for greed or hunger
A brotherhood of man
*Imagine all the people **sharing all the world**, you*
You may say I'm a dreamer
But I'm not the only one
I hope someday you'll join us
And the world will be as one

Now, before you get all teary-eyed and sniffly; pause to consider if these noble-sounding platitudes are realistic. Are we to believe that a ruthless Mafia of liars, control freaks and money junkies, upon achieving the ultimate objective of Global domination, will suddenly convert to benevolent and wise rulers of humanity?

"NEW WORLD ORDER TO SAVE EARTH" *-- Former UK Prime Minister*
Gordon Brown   // UN Calls for One World Currency

The **New World Order** isn't about a bunch of diverse kids standing in an open field holding hands and singing: *"I'd like to buy the world a Coke."* That's the mushy soft-hearted bait. The system is about power, with the elites at the top, and the rest of us tax-paying, city-dwelling, auto-less worker bees on the lower levels – toiling away to support the rotten structure as we enjoy our basic "three hots and a cot" and little else. Forget John Lennon. Think George Orwell!

Professor **Frederick Lindemann** *(1886-1957)* was Winston Churchill's top advisor and most likely, his handler. Though little known to the public, his influence over Churchill was immense. Lindemann held that a small circle of elites should run the world, resulting in a stable society, *"led by supermen and served by helots."* **(28)** He believed that science could yield a race of humans blessed with *"the mental makeup of the worker bee."* **(29)** Of course, Lindemann's class of political leaders and fellow mad scientists won't be among the "worker bees." That's for sure.

For those of you who believe that man was born to be free and independent, and governed by wise and honest leaders who live and breathe for the happiness, fulfilled potential, and cultural advancement of their people, this New World Order of "worker bees" should be resisted with everything you've got. Worker bees literally work themselves to death and with nothing to show for it at the end – not unlike more and more Americans and Europeans are today!

Who are you people to lord over us by trickery and deceit? Take your bloody Globalist "beehive" and stick it where the sun don't shine, with the angry bees still in it -- **Mr. Lindemann**, and **Mr. Gore**, and **Mr. Obongo**, and **Mr. DiCaprio**, and **Mr. Pope**, and **Ms. Merkel**, and **Prince Charles**, and **Mr. Gates**, and **Mr. Soros**, and **Mr. Rothschild**, and **Mr. Zuckerberg**, and **Mr. Hawking** and **Mr. Branson,** and **Mr. Sulzberger,** and **Mr. Degrasse-Tyson, and Mr. Nye** et al.

Instead of putting "Climate Deniers" in jail, as Bill Nye has actually suggested, **(30)** it is the warmists who ought to be investigated – the innocent dupes to be forgiven, but the conspirators *(and there are many)* to be brought up on charges of high treason. It is the elite members of the "save the planet" Globalist Mafia who are the REAL "bogeymen."

*1. Professor Lindemann wanted a world system of "worker bees" tightly controlled by Globalist Mafia elite. 2. Billionaire mobsters **Warren Buffett, Bill Gates** and **George Soros** are all "in on it." 3. The creepy eyeball and pyramid on the back of a $1 bill says it all: "**Novus Ordo Seclorum**" – a New Order for the Ages. That's what the GW/CC hoax is all about.*

CONCLUSION & CALL TO ACTION

What is to be done about this criminal conspiracy against the people of the world? It all starts with bringing this critical and accurate information to those who have been fooled -- *especially* High School and University students who are getting hammered with this Globalist garbage.

It is a daunting task, given the immense interconnected political power and extreme wealth that we are up against. But, as the warmists like to say: "Think globally. Act locally."

Why not start your own public education campaign by ordering a few extra copies of **Climate Bogeyman** and gifting them to friends and family members – particularly school and college kids.

You may even want to address a copy to the "Environmental Science" department at a local High School. Include a nice note and do it anonymously if you feel the need to do so. By educating an innocently misguided educator, you can save many young minds at a time. As the hideous green-haired freak, "Captain Planet" used to say at the close of each cartoon: "The power is yours!"

RESOURCES: RECOMMENDED READING & VIEWING

WEBSITES

CLIMATE DEPOT

Marc Morano's ClimateDepot.com

Anthony Watt's WattsUpWithThat.com

187

FILMS

The Climate Hustle
www.ClimateHustle.org

YouTube: The Great Global
Warming Swindle

YouTube or Amazon: The
Greenhouse Conspiracy

NotEvilJustWrong.com

BOOKS

Climategate, By Brian Sussman

The Hockey Stick Illusion, By A.W. Montford

The Climate Caper, By Garth Paltridge

The Real Global Warming Disaster, By Christopher Booker

The Geatest Hoax, By Senator James Inhofe

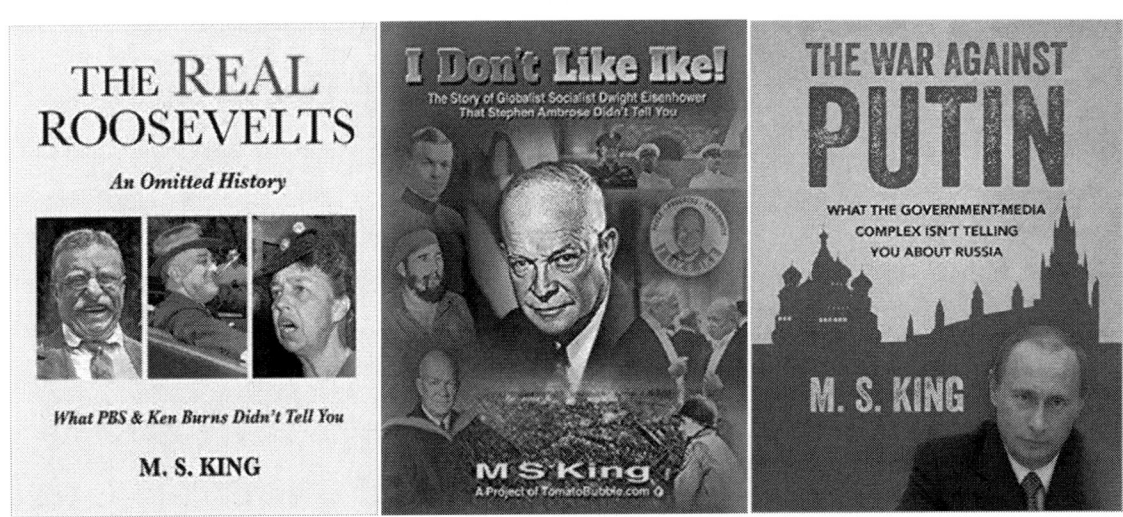

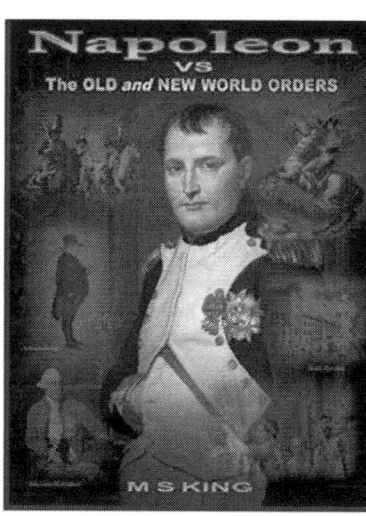  

TomatoBubble.com

The Mind-Altering Internet Classics of Alternative History, Economics, Philosophy and Current Events

FOOTNOTES

The innovative use of Internet Search terms as footnotes allows the curious reader to trace the quotes and hard facts of this book to multiple credible sources, instead of just one.

SECTION 1

1. Internet Search: The Guardian, September 23, 2014, Leonardo DiCaprio at the UN

2. Internet Search: Greenhouse Conspiracy PBS I'm not sure useful include point of view

3. Internet Search / YouTube: Captain Planet CO2 brainwashing

4. Ibid

5. Internet Search: Inconvenient Truth for Kids: Best-seller available kid-friendly version

6. Internet Search / You Tube: Gingrich Pelosi on love seat global warming

7. Internet Search Obama Berlin speech 2008

8. Internet Search: Obama San Francisico Chronicle interview 2008

9. Ibid

10. Internet Search: Prince Charles: 100 months to avert catastrophic climate change

11. Internet Search: Daily Mirror: Prince Charles 30 Years to Save Planet Catastrophe

12. Internet Search: Hansen Obama has four years to save the world

13. Internet Search: text of obama speech copenhagen climate summit

14. Internet Search: Pope Francis It is man who continuously slaps down nature

15. Internet Search: Pope urgent to develop policy in coming years reduce carbon dioxide

16. Internet Search: Dicaprio we went corner of globe document impacts climate change

SECTION 2

1. Internet Search: Tyson if anyone says it is just a theory missing education

2. Internet Search: tim wirth We've got to ride the global warming issue

3. Internet Search: tesla Today's scientists have substituted mathematics for experiments

4. Internet Search: Yale University: How Long Can Oceans Continue To Absorb Earth's Excess Heat?

5. Internet Search: New York Times: Ocean Life Faces Mass Extinction, Broad Study Says

6. Internet Search: ABC Global warming: Australian scientists strong winds Pacific pause temperatures

7. Internet Search: Ted Cruz's Favorite Argument about Climate Change Just Got Weaker

8. Internet Search: Phys.org revamped Satellite Data Shows No Pause in Global Warming

9. Internet Search: Scott Shackleton logbooks prove Antarctic ice not shrinking 100 years after

10. Internet Search: NASA Study: Mass Gains of Antarctic Ice Sheet Greater than Losses

11. Internet Search: Washington Times: Irony alert: Global Warmists get stuck in ice

12. Internet Search: daily mail Climate change experts trying explain why sea ice Antarctica expanding.

13. Internet Search: Forbes Magazine Updated NASA Data: Global Warming Not Causing Polar Retreat

14. Internet Search: Daily Caller Global Warming Expedition Stopped In Its Tracks By Arctic Sea Ice

15. Internet Search: 1975 – Newsweek: The Cooling World (Peter Gwynne, April 28, 1975)

16. Internet Search: YouTube: nimoy in search of the coming ice age episode

17. Internet Search: ibid

18. Internet Search: New York Times, January 8, 1975, stephen Schneider

19. Internet Search: forbes july 17 2012 That Scientific Global Warming Consensus...Not!

20. Internet Search: petitionproject.org

21. Internet Search: July 23, 2013, Nottingham What's behind battle of received wisdoms

22. Internet Search: Judith Curry resigns directions approved by a politicized academic establishment

23. Internet Search: Space.com *(May 26, 2016)*: Red Planet Heats Up: Ice Age Ending on Mars

24. Internet Search: EPA document supports 3% of atmospheric carbon dioxide attributable to human sources

https://wattsupwiththat.com/2014/07/29/epa-document-supports-3-of-atmospheric-carbon-dioxide-is-attributable-to-human-sources/

25. Internet Search: ibid

26. Internet Search: The New York Slimes: Termite Gas Exceeds Smokestack Pollution

27. Internet Search: ScienceDaily.com winds of change wind patterns have shifted

28. Internet Search: Daily Mail Changing wind patterns, not global warming, causing temperatures to rise on America's West Coast

29. Internet Search: scientificamerican.com how do volcanoes affect world climate

30. Internet Search: Principia Scientific: New Study: Solar & Cosmic Rays Impact Climate

31. Internet Search: thames river frost fair

32. Internet Search: New York Times Story of Viking Colonies' Icy 'Pompeii' Unfolds From Greenland

33. Internet Search: hawking we are close to tipping point where global warming becomes irreversible

34. Internet Search: NPR Disappearing Montana Glaciers a 'Bellwether' Of Melting To Come?

35. Internet Search: Live Science Why Asia's Glaciers Are Mysteriously Expanding, Not Melting

36. Internet Search: Phys.org Explaining New Zealand's unusual growing glaciers

37. Internet Search: New York Times the Marshall Islands Are Disappearing

38. Internet Search: Inverse.com Six Pacific Islands Have Already Disappeared as Sea Levels Rise

39. Internet Search: The London Telegraph: Low-Lying Pacific islands 'growing not shrinking' due to climate change

40. Internet Search: UK Daily Mail The Island That 'Grew Back': Pacific Isle That Disappeared After Devastating Typhoon Reappears 100 Years After Its Destruction

41. Internet Search: New Scientist.com Small atoll Islands May Grow, Not Sink As Sea-Levels Rise

42. Internet Search: New York Times A Drought in Australia, a Global Shortage of Rice

43. Internet Search: New York California Drought Is Made Worse by Global Warming, Scientists Say

44. Internet Search: New York Times: Pray for Shade: Heat Wave Sets a Record in India ....

45. Internet Search: New York Times Too Hot to Fly? Climate Change May Take Toll on Air Travel

46. Internet Search: telegraph.co.uk Christopher booker rise of sea levels greatest lie ever told

47. Internet Search: Dr. Morner *the sea is not rising. It has not risen in 50 years*

48. Internet Search: Dr. Morner The late 20th Century sea-level rise lacks any sign of acceleration.

49. Internet Search: Dr. Morner frighten scientists if they say climate not changing lose research grants

50. Internet Search: Dr. Morner level rise does not exist in observational data, only in computer modeling

51. Internet Search: New York Times the Climate Refugees of the Arctic

52. Internet Search: ibid

53. Internet Search: Daily Express Polar bear populations recovering despite the climate change warnings

54. Internet Search: PolarBearScience.com Survey Svalbard polar bear numbers increased 42% 11 years

55. Internet Search: ibid

56. Internet Search: National Geographic Longest Polar Bear Swim Recorded—426 Miles Straight

57. Internet Search: www.bearlife.org/baby-polar-bears.html

58. Internet Search: ibid

59. Internet Search: 1912: A winter of record cold

60. Internet Search: Global Warming Bombshell evidence linking human activity to climate change poor mathematics

61. Internet Search: Victor Marshall The Lies We Are Told About Iraqhttp://articles.latimes.com/2003/jan/05/opinion/op-marshall5

62. Internet Search: business insider NASA scientists dispute climate change

63. Internet Search: exposed: world leaders duped investing billions manipulated global warming data

64. Internet Search: *New York Times* World Leaders Move Forward on Climate Change, Without U.S.

SECTION 3

1. Internet Search: plato *see* beyond the shadows and lies of their culture

2. Internet Search: New York Times Too Hot to Fly? Climate Change May Take Toll on Air Travel

3. Internet Search: The Guardian Revealed: first mammal species wiped out by climate change

4. Internet Search: The Guardian Devastated: scientists too late to breed mammal lost to climate change

5. Internet Search: Environmentalist Blogger,Michelle Nijhuis Who Killed the Bramble Cay Melomys

6. Internet Search: Washington Post Bramble Cay melomys mammal claimed climate change report says

7. Internet Search: New York Times Australian Rodent First Mammal Made Extinct by Climate Change

8. Internet Search: New York Times Australian Rodent First Mammal Made Extinct by Climate Change

9. Internet Search: National Wildlife Federation: Global Warming and Drought

10. Internet Search: The Guardian (UK): Global Warming is Increasing Rainfall Rates

11 Internet Search: National Wildlife Federation: Global Warming and Heat Waves

12. Internet Search: Scientific American: Global Warming Can Mean Harsher Winter Weather

13. Internet Search: Weather Underground: Arctic Sea Ice Decline

14. Internet Search: Nature: Ocean Warming May Be Major Driver of Sea-Ice Expansion in the Antarctic

15. Internet Search: Scientific American: Climate Change May Mean Slower Winds

16. Internet Search: Live Science: Global Warming Weakens Trade Winds

17. Internet Search: ABC (Australia): Global Warming: Australian scientists say strong winds in Pacific Behind pause in Rising Temperatures

18. Internet Search: University of California at Santa Cruz: Stronger coastal winds due to climate change may have far-reaching effects

19. Internet Search: National Geographic: The Big Thaw: As climate warms, how much, and how quickly, will Earth's glaciers melt?

20. Internet Search: National Geographic: Some Glaciers Growing Due to Climate Change

21. Internet Search: United Nations University: Atoll islands and climate change: disappearing States?

22. Internet Search: Telegraph (UK): Pacific Islands Growing, Not Shrinking Due to Climate Change

23. Internet Search: UK Independent: Global Warming is Causing More Hurricanes

24. Internet Search: NASA: In a Warming World, the Storms May Be Fewer But Stronger

25. Internet Search: www.numberwatch.co.uk/warmlist.htm

26. Internet Search: http://hockeyschtick.blogspot.com/2013/06/climate-scientist-dr-murry-salby.html

27. Internet Search: BirdWatchingDaily.com New study estimates 573,000 birds died at wind farms

28. Internet Search: DeGrasse Tyson: "Philosophy not productive contributor to understanding natural world …. It can really mess you up.

29. Internet Search: Hawking: Philosophy is dead. Philosophers have not kept up with modern science.

30. Internet Search: Nye Philosophy important for while start arguing circle humans made up philosophy

SECTION 4

1. Internet Search: obama under my plan of a cap and trade system

2. Internet Search: This excerpt is from an E-mail that a reader from Iowa sent in to TomatoBubble.com. Mark knows a lot about the windmill scam in Iowa and lives close to some of them. I can personally vouch for Mark and the authenticity of his home's proximity to Iowa windmills.

3. Internet Search: New American: 14,000 Idle Wind Turbines Testament Failed Energy Policies

4. Internet Search: Newsweek: What is the true cost of wind power?

5. Internet Search: Wind Power Monthly Annual blade failures estimated at around 3,800

6. Internet Search: forbes The Myth Of Wind And Solar 'Capacity'

7. Internet Search: national review The Biofuel Scam Is Worse than Solyndra

8. Internet Search: obama Under my *plan of a* cap and trade system

9. Internet Search: ibid

10. Internet Search: ibid

11. Internet Search: forbes clement Cap-And-Trade Is Fraught With Fraud

12. Internet Search: report shows zero statistically different result business as usual Susan Satter

13. Internet Search: We have looked at most elements of smart grid for 20 years we have never come up with estimates that make it pay John Rowe,

14. Internet Search: No net economic benefit to ratepayers Bill Schuette,

15. Internet Search: Consumers Digest Smart-meter conversion represents boondoggle being foisted on consumers politically influential companies

16. Internet Search: http://www.smartmetereducationnetwork.com/smart-meter-costs.php

17. Internet Search: daily caller EPA Regulations To Cause Double-Digit Electricity Price

18. Internet Search: daily caller obama kept his promise 83000 coal jobs lost

19. Internet Search: James Shikwati killing the African dream

SECTION 5

1. Internet Search: obama berlin speech 2008

2. Internet Search: edith Kermit Roosevelt Elite Clique Holds Power in U.S

3. Internet Search: chester ward once the ruling members of the CFR have decided that the U.S

4. Internet Search: bill Clinton carroll quigley

5. Internet Search: Quigley There does exist, and has existed for a generation

6. Internet Search: edith Kermit Roosevelt Elite Clique Holds Power in U.S

7. Internet Search: chester ward submergence sovereignty national independence **one-world** government

8. Internet Search: Quigley the powers of financial capitalism had another far-reaching aim

9. Internet Search: Strobe Talbott In next century, nations as we know it will be obsolete

10. Internet Search: Walter Cronkite We Americans are going to have to yield up our sovereignty.

11. Internet Search: Hillary *Clinton* congratulations Walter receiving World Federalist Global Governance Award

12. Internet Search: Einstein a Federal organization of the nations of the world

13. Internet Search: pope francis there is urgent need of a true world political authority

14. Internet Search: Barack Obama: All nations must come together to build a stronger, global regime

15. Internet Search: Senator Barry Goldwater: "David Rockefeller's newest international cabal,

16. Internet Search: Larry MacDonald: drive of Rockefellers and allies to create one-world government

17. Internet Search: Congressman Ron Paul We're only supposed to talk about internationalism, globalism, one-world government

18. Internet Search: putin The UniPolar world refers to a world in which there is one master

19. Internet Search: Trump: We will no longer surrender country or people to false song of Globalism

20. Internet Search: Bernays conscious and intelligent manipulation of the organized habits and opinions

21. Internet Search: hitler a last appeal to reason

22. war cabinet crisis Churchill halifax

23. Internet Search: walter isaacson and evan thomas --- the wise men ...p.410

24. Internet Search: Shadows of Power, by James Perloff – citing charles l. mee jr. the marshall plan, simon & schuster, 1984, p.234

25. Internet Search: The Chicago Tribune (February 10, 1952): Brother Milton Worry Aids Eisenhower

26. Internet Search: *General* Wesley clark seven countries five years

27. Internet Search: We cannot wait for the smoking gun in the form of a mushroom cloud

28. Internet Search: Lindemann led by supermen and served by helots

29. Internet Search: Lindemann the mental makeup of the worker bee

30. Internet Search: washington times Bill Nye open criminal charges and jail climate change dissenters

54144428R00111

Made in the
USA
Lexington, KY